TABLE OF CONTENTS

i

INTRODUCTION

> *The next Pearl Harbor we confront could very well be a cyber attack that cripples our power systems, our grid, our security systems, our financial systems, our governmental systems. As a result, I think we have to aggressively be able to counter that. It is going to take both defensive measures as well as aggressive measures to deal with it.*
>
> \- Secretary of Defense Leon Panetta

The Fifth Domain

It has been about one hundred years since warfare expanded into the third dimension. Since then, war in the air has become commonplace and has grown into the fourth dimension, the ultimate high ground, in outer space. Now, in consonance with the more recent information revolution, a new venue for warfare has emerged—the cyber domain, one characterized by former Air Force Chief of Staff General Larry D. Welch as the "fifth operational domain."[1] And much like the early period of airpower almost a century ago, a crowd of prognosticators have put forth a confusing array of theories, each seeking to explain war, security, defense, and victory in cyberspace. Unfortunately, as with many of the early airpower theories, these emerging cyberspace theories are, more often than not, ill-informed, incoherent, and ignorant of some of the most important lessons from history.

None of that would be significant except that thousands of cyberspace attacks occur each day and the likelihood of a significant incident in the near future is high. At his Senate confirmation hearing, Defense Secretary Leon Panetta emphasized that without the creation of a coherent strategy to deter attacks through cyberspace, the U.S. is

---

[1] Land, sea, air, and space being the other four domains. Gen. Larry D. Welch USAF (Ret.), "Cyberspace-The Fifth Operational Domain," *IDA Research Notes*, Summer 2011, https://www.ida.org/upload/research%20notes/researchnotessummer2011.pdf (accessed August 8 2011).

at risk of another catastrophic surprise attack on par with Pearl Harbor or 9/11.[2] Sadly, as with the emergence of previous "game changing" developments in powered flight and atomic weaponry, the development of a coherent cyberstrategy has been sluggish. This delay often emanates from the perception that cyberspace represents such a pervasive revolution in the conduct of warfare that successful deterrence strategies of the past are not applicable. Authors such as RAND's Martin Libicki continue to insist that, "attempts to transfer policy constructs from other forms of warfare [to cyberspace] will not only fail but also hinder policy and planning."[3] What Libicki and others apparently fail to appreciate is that the U.S. can only achieve effective cyberspace deterrence by applying the principles, lessons, and knowledge gleaned from the history of warfare in other domains. In formulating the foundation of airpower theory in 1921, Giulio Douhet, like Libicki, made similarly short-sighted declarations that "clinging to the past will teach us nothing useful for the future, for that future will be radically different from anything that has gone before."[4] Ninety years later, it is clear that the airplane did radically change the character of warfare, but the nature and principles of war still apply despite distinct characteristics of air, land, sea and space employment. The same logic holds true in capturing the significance of the role of cyberspace as it relates to warfare. The thesis of this paper is that analysis of deterrence during key periods of history reveals a significant practical foundation for developing strategies to prevent attacks in the new domain of cyberspace. President Obama emphasized the importance of this approach in his introduction of the new Department of Defense (DoD) strategy, *Sustaining Global*

---

[2] Tony Capaccio, "CIA Director Panetta Gets Senate Nod as Next Defense Secretary," *Bloomberg Businessweek*, June 22, 2011, http://www.businessweek.com/news/2011-06-22/cia-director-panetta-gets-senate-nod-as-next-defense-secretary.html (accessed October 16, 2011).

[3] Martin C. Libicki, *Cyberdeterrence and Cyberwar*, (California: RAND, 2011), xiii.

[4] Guilio Douhet, *The Command of the Air* (New York: Coward-McCann, Inc., 1942), 26.

Leadership: Priorities for the 21st Century Defense. The President eloquently stated: "going forward, we will also remember the lessons of history and avoid repeating the mistakes of the past when our military was left ill-prepared for the future."[5] He also emphasized that "prevailing in all domains, including cyber" is a capability "critical to future success" of the United States.[6]

The Known

While commanding the 18th Aggressors Squadron at Eielson Air Force Base (AFB), it became clear to this researcher that the unit's mission was to place technically proficient pilots in a realistic combat environment so that when in actual combat they would innately see the enemy's tactics as not dramatically new, but analogous to their earlier exposure. The human mind looks at a stored model, or paradigm, to understand what it "sees." It was critical to train pilots that combat, in a different environment, was not something altogether new, but instead an application of what they knew to a different situation and threat. Analogous awareness, training, and developed thinking skills are the key to utilizing past experience and wisdom as a basis for building the most effective strategies for future challenges. To assume the posture of revolutionary change leads to the conclusion that those trained and experienced in the old methods are not helpful, needed or able to address the new challenges. Cyberspace certainly offers some new challenges, but the most effective and efficient solution comes from seeing the patterns in the underlying issues and establishing defense and deterrence strategies based on a proven historical foundation.

---

[5] U.S. Department of Defense, *Sustaining U.S. Global Leadership: Priorities for the 21st Century Defense* (Washington DC: Department of Defense, January 2012), 1.
[6] Ibid.

Military leaders study, employ, and adapt the writing of Clausewitz, Mahan, and Douhet as well as other great military practitioners, using their historically based principles to highlight the dominant land, sea and air capabilities. These capabilities play a critical factor in the most fundamental strategic decision, whether or not to go to war at all. The idea of deterring aggression is as old as warfare and involves much more than the military instrument of power. Sun Tzu noted in his writings from roughly 500 B.C. that "to subdue the enemy without fighting is the acme of skill."[7] The strategy of military deterrence is thousands of years old, yet the rapid development of deterrence theory is primarily associated with the period following World War II when the U.S. and the Soviet Union reached a strategic nuclear stalemate of Mutually Assured Destruction.[8] Deterrence theories developed and refined during this period, under the auspices of Eisenhower's New Look and Kennedy's Flexible Response policies, offer useful insights into the creation of effective deterrence strategies for all domains, including cyberspace. Moreover, the efforts to deter militant leaders such as Muammar Qaddafi and Saddam Hussein from using Weapons of Mass Destruction (Chemical, Biological, Radiological and Nuclear) also provide a launching pad for designing an overall deterrence policy for cyberspace. Finally, the end of the Cold War and the destruction of the World Trade Center on September 11, 2001 widened America's deterrence focus from simply peer competitors to terrorist threats and rogue states with an emphasis on proactive destruction of their capabilities to wage war.[9] In other words, a brief historical analysis of deterrence

---

[7] Samuel B. Griffith, *Sun Tzu: The Art of War* (London: Oxford University Press, 1963), 77.
[8] A.J.C. Edwards, *Nuclear Weapons, the Balance of Terror, and the Quest for Peace* (London: MacMillan, 1986), 3.
[9] U.S. Government, *The National Security Strategy of the United States of America* (Washington D.C.: September 2002), 14.

during these periods reveals a significant practical foundation for cyberspace strategy development.

The Way Ahead

As Shakespeare wrote in *The Tempest*, "What's past is prologue." The U.S. needs an effective, comprehensive, and integrated cyberspace deterrence strategy. This plan must include defense of critical government and civilian networks from the full range of cyberspace attacks. Absent such an approach, the U.S. risks the real possibility of catastrophic damage from those seeking to do harm in and through this new warfighting domain. This challenge, while radical in its speed, precision, and effect, is nonetheless similar to others faced throughout military history. Today, existing threats in cyberspace are the most pressing and compelling test of our thinking.

The Joint Staff (J-5) requested additional academic research on "Deterence in Cyberspace" through the Prospective Topics Research Database on 7 August 2010. Although there is a large amount of contemporary literature on cyberspace warfare and deterrence, joint doctrine for cyberspace operations or deterrence has not yet been developed. This research and the principles for deterring malicious activity in cyberspace aim to provide a basic vector for follow-on doctrine development.

Research Approach

Chapter 1 is a discussion of the important definitions that underlie both the challenge and the solution for deterrence in cyberspace. Chapter 2 provides the historical context and principles of deterrence in other domains. Chapter 3 defines the development of an effective deterrence strategy for cyberspace by applying basic principles gleaned from history to the cyberspace domain. Chapter 4 makes further progress toward an effective, efficient, and successful cyberspace deterrence strategy by identifying the current knowledge shortfalls that require further research.

CHAPTER 1: FOUNDATIONS

> *There is something more important than any ultimate weapon. That is the ultimate position — the position of total control over Earth that lies somewhere out in space. That is . . . the distant future, though not so distant as we may have thought. Whoever gains that ultimate position gains control, total control, over the Earth, for the purposes of tyranny or for the service of freedom.*
>
> - Senator Lyndon B. Johnson, 1958

The term cyberspace first appeared in the short story "Burning Chrome" by Canadian science fiction writer William Gibson in 1982.[1] The term caught on rapidly as it seemed to characterize perfectly the World Wide Web and the meteoric rise of network capabilities. The first widespread military use of the term resulted from the Air Force adding cyberspace to its mission statement in 2005: to *fly, fight, and win…in air, space and cyberspace.*[2] A casual search of the internet now reveals over twenty definitions for cyberspace. This chapter serves to establish a common definition for cyberspace as well as potentially contentious terminology such as *domain*, the *information environment, information technology infrastructures* and *deterrence*. In addition, a broad discussion of potential cyberspace threats and current policy documents serves to identify shortfalls within the current cyberspace deterrence approach that the principles established through historical analysis serve to assuage.

Cyberspace

The broad use of the term "cyberspace," coupled with the lack of a common lexicon, prompted the Deputy Secretary of Defense, Gordon England, to publish a memo

---

[1] William Gibson, *Burning Chrome* (New York: Omni, 1982).
[2] MSgt Mitch Gettle, "Air Force Realeases New Mission Statement," December 8, 2005, http://www.af.mil/news/story.asp?id=123013440 (accessed Novemeber 26, 2011).

to all government agencies on 12 May 2008 solidifying the DoD definition as "a global domain within the information environment consisting of the interdependent network of information technology infrastructures, including the internet, telecommunication networks, computer systems, and embedded processors and controllers."[3] Published in 2011, Joint Publication 1-02, *The Department of Defense Dictionary of Military and Associated Terms*, now defines cyberspace within the military vocabulary with precisely the same definition.[4] Previous to Mr. England's memo, debates regarding cyberspace tended to start and end with uncertainty regarding a common definition. The ubiquity of cyberspace makes a common definition essential and, with a firm understanding of what cyberspace represents, a logical discussion of the capabilities and limitations can advance. Before moving on, however, key components of this definition need to be discussed, beginning with the idea of a domain.

Domain

The definition above characterizes cyberspace as a "global domain." What does the term domain mean and why is it significant to cyberspace? The word "domain" is not defined in Joint Publications but, according to *The American Heritage Dictionary*, the most appropriate definition in this context is a "territory over which rule or control is exercised."[5] The boundaries of cyberspace, though vast, are defined by the physical hardware architecture of the network, terminating with the end user. These boundaries define a *territory* over which various private and government agencies exercise *control*.

---

[3] U.S. Department of Defense, *The Definition of "Cyberspace,"* Gordon England, Memorandum, Deputy Secretary of Defense (Washington D.C., 2008).

[4] U.S. Joint Chiefs of Staff, *Department of Defense Dictionary of Military and Associated Terms*, Joint Publication 1-02 (Washington DC: Department of Defense, October 2011), 86.

[5] American Heritage Dictionary of the English Language (Boston: Houghton Mufflin Company, 2000), 233.

Given the characterization as both a territory and one able to be controlled, cyberspace classifies as a domain. The *Department of Defense (DoD) Strategy for Operating in Cyberspace*, published in July 2011, reinforces this concept and outlines as its first strategic initiative the intent that "DoD will treat cyberspace as an operational domain to organize, train, and equip so that DoD can take full advantage of cyberspace's potential."[6] In addition, Deputy Defense Secretary William J. Lynn III emphasized that "cyberspace is a warfighting domain: like air, sea, land, and space, we're going to have to treat cyberspace as an arena where we need to defend our networks and to be able to operate freely."[7] Given a common definition, it is now important to analyze how cyberspace exists within the "information environment."

Information Environment

One of the most essential elements in understanding the cyberspace domain is first to recognize that it exists within a much broader area known as the information environment. Some would argue that information represents a warfighting domain in itself and, in fact, JP 3-0, *Joint Operations*, first characterized information as a separate domain.[8] A failure to reach a consensus within the Joint community led to the characterization of information as an "environment" rather than a domain.[9] The root cause of this change is the lack of defined boundaries on information. An environment is "the aggregate of surrounding things, conditions, or influences" and thus deemed a more

---

[6] U.S. Department of Defense, *Department of Defense Strategy for Operating in Cyberspace* (Washington DC: Department of Defense, July 2011), 5.

[7] Ibid.

[8] U.S. Joint Chiefs of Staff, *Joint Operations*, Joint Publication 3-0 (Washington DC: Department of Defense, 17 September 2006), iv.

[9] Ibid., 2.

appropriate definition of the borderless information cloud that surrounds us.[10] Joint

Publication 3-13, *Information Operations*, uses the following definition to describe this

environment: "The information environment is the aggregate of individuals,

organizations, and systems that collect, process, disseminate, or act on information. The

information environment is made up of three interrelated dimensions: physical,

informational, and cognitive."[11] Cyberspace is a crucial component of this environment

and networked systems, used to transport data, are a part of every aspect of our daily

lives. Each day it becomes more evident that the U.S. is increasingly reliant on

networked systems to function at even the most basic levels. This reliance brings with it

a rapid increase in power, access, and capability that network security efforts simply

cannot match. This failure is commonly associated with inadequate defenses, but is more

closely tied to the lack of a suitable deterrence strategy. Deterrence in any domain is an

Information Operation and cyberspace deterrence denotes preventing action within the

cyberspace domain by affecting the decision making process of the would-be attacker.

Effective strategy utilizes a host of cross-domain actions including diplomatic,

information, military, and economic instruments of power that can deter attacks. The

ultimate aim is to convince an aggressor that taking a specific action within cyberspace,

stealing information about a military weapon system, for example, will bring too high of

a cost relative to the expected gain. It is, therefore, vital to look at the range of threats

within cyberspace that an organization must deter in order to formulate the appropriate

defense. The best defensive strategy is to determine the path to prevent attacks from

---

[10] American Heritage Dictionary, 345.

[11] U.S. Joint Chiefs of Staff, *Information Operations*, (Washington DC: Department of Defense, February 2006), x.

happening in the first place. Unfortunately, many senior leaders shy away from this problem because they view cyberspace as something altogether new.[12]

Information Technology Infrastructures

The definition of cyberspace specifies that it consists of an "interdependent network of information technology infrastructures, including the internet, telecommunication networks, computer systems, and embedded processors and controllers."[13] This terminology further emphasizes the physical boundary associated with the cyberspace domain. Cyberspace is the only man-made domain and consists of the physical components that enable data to flow from one computer system to another. Unlike air, sea, land and space, people designed and built the components that comprise this new and far-reaching domain of warfare. This territory includes the technology to transmit data, or "information technology infrastructures," as well as the computer processors and logic controllers that compose the inner workings of computers, routers, switches, hubs, servers, and the host of other electronic components that enable computer processing and networking in general. In the context of a conflict, however, the nature of a domain has little meaning without the actual tools of waging war. Bombs, bullets, missiles or computer viruses characterize a conflict much more than the domain or medium within which the battle is fought. The next section demonstrates that deterring the employment of these warfighting tools has commonality irrespective of the boundaries of land, air, sea, space or cyberspace.

---

[12] Gen Michael V. Hayden, "The Future of Things Cyber," *Strategic Studies Quarterly* 5, no. 1 (Spring 2011): 3.
[13] Ibid.

Deterrence

Joint Publication (JP) 1 states, "deterrence helps prevent adversary action through the presentation of a credible threat of counteraction."[14] The key components of this definition are "helps prevent" and "credible threat." The lack of a nuclear exchange during the Cold War is the best example of a successful deterrence strategy, but years of postmortem study reveal how dangerously close the U.S. and Soviet Union came to war during the Cuban Missile Crisis. President Kennedy emphasized this point after the crises ended by saying that he thought the odds of war were "between one out of three and even."[15] Ultimately, no matter how close war looms, prevention of a conflict is effective deterrence by any measure. There is no definitive way to know whether a deterrence strategy will be effective, however, primarily because it is difficult to predict the cognitive logic of the adversary given what each might consider a "credible threat." If an adversary does not perceive significant repercussion for malicious actions, then deterrence has failed. Deterrence, at the most basic level, exists in all aspects of life as everyday citizens assess risks in engaging in a particular action relative to the reward. If a person speeds along the highway, there is the threat of being ticketed, but a driver may perceive that the risk may be worth taking if they are running late for an important event. Deterrence has two distinct elements--denial and the threat of punishment. Denial involves convincing the adversary that they will be unlikely to achieve their objective, normally due to a strongly defended target. The fear of punishment or reprisal involves threats to destroy something of great value to the adversary as the result of their own

---

[14] U.S. Joint Chiefs of Staff, *Doctrine for the Armed forces of the United States*, Joint Publication 1 (Washington D.C.: Joint Chiefs of Staff, March 20, 2009), 1-16.
[15] Richard E. Neustadt and Ernest R. May, *Thinking in Time*, (New York: The Free Press, 1986), 13.

attack.[16] As stated in the most recent DoD strategic guidance, "credible deterrence results from both the capabilities to deny an aggressor the prospect of achieving his objectives and from the complementary capability to impose unacceptable costs on the aggressor."[17] It is important to understand that deterrence is not solely reliant on prevention through denial and fear, but can also be achieved through other means. Incentives for favorable behavior also serve as an effective deterrent as demonstrated by the Six-Party Agreement between the U.S., China, Russia, Japan, South Korea, and North Korea in February 2007.[18] This agreement guaranteed heavy fuel oil and food aid to North Korea in exchange for freezing and allowing inspections of activities within its Yongbyon nuclear facility, key to nuclear weapons development.[19] Therefore, deterrence can be used in the common context of prevention through fear, but also in a more general application of incentivizing favorable or non-belligerent behavior.

Cyberspace Threats

The 2010 *National Security Strategy* emphasizes that "cybersecurity threats represent one of the most serious national security, public safety and economic challenges we face as a nation."[20] Cyberspace threats run the gamut from simple criminal activity, such as identity theft, to strategic effects on par with nuclear weapons. Until recently, malicious cyberspace activities were primarily associated with criminal undertakings. Russia's August 2008 campaign against Georgia was a watershed event in that it marks the first widespread use of cyberspace forces in conjunction with a

---

[16] Neustadt and May, *Thinking in Time*, 18.
[17] *Sustaining U.S. Global Leadership: Priorities for the 21st Century Defense*, 4.
[18] Dick K. Nanto and Emma Chanlett-Avery, *North Korea: Economic Leverage and Policy Analysis* (Washington D.C.: Congressional Research Service, January 2010), 23.
[19] Ibid.
[20] U.S. President, *The National Security Strategy of the United States of America* (Washington D.C.: Government Printing Office, May 2010), 27.

conventional force.[21] Cloaked in the façade of the Russian Business Network (RBN), the attacks affected thirty-eight websites, including the Georgian President, the Ministry of Foreign Affairs, the National Bank, the Parliament, the Supreme Court, and the embassies of the U.S., and United Kingdom in Georgia.[22] As computing technology and global networking advance, these types of attacks will no longer be limited just to nation-states. Indeed, even today non-state actors, individuals, and corporations engage in both cyberspace attacks and espionage.[23] With time, the cyberspace capabilities devised by both nations and terrorists will potentially have destructive impact equal to a major natural disaster or nuclear detonation.

In late 2009, the STUXNET worm affected 984 centrifuges used to enrich Uranium for the production of nuclear weapons in Iran.[24] This worm targeted the Siemens Process Control System (PCS) and so heavily damaged equipment vital to nuclear weapons production that Iran's program will take years to recover.[25] Although not specifically attributed to the U.S. or Israel, STUXNET exhibited the capabilities born out of years of study and testing tied to only a handful of countries with highly advanced offensive cyberspace capabilities. These same types of vulnerabilities, however, exist within the broader class of critical infrastructure systems, known as Supervisory Control and Data Acquisition (SCADA) systems. The SCADA networks monitor and control industrial or infrastructure based computer systems responsible for a variety of processes such as power generation, manufacturing as well as oil and gas distribution pipelines.

---

[21] Ariel Cohen and Robert E. Hamilton, *The Russian Military and the Georgia War: Lessons and Implications* (Carlisle: Strategic Studies Institute, 2011), 54.

[22] Ibid., 44-45.

[23] McAfee Foundstone Professional Services and McAfee Labs, "Global Energy Attacks: Night Dragon," *White Paper*, February 10, 2011, 3-4, 18.

[24] William J. Broad, John Markoff and David E. Sanger, "Isreali Test on Worm Called Crucial in Iran Nuclear Delay," *The New York Times*, January 15, 2011.

[25] Ibid.

The vulnerability of SCADA systems or its components, such as the Siemens PCS, exposes the possibility that an adversary could achieve devastating effects on domestic nuclear power systems, the domestic power grid, and a host of other such systems that millions simply take for granted.

The Aurora experiment in 2007 is yet another demonstration of how actions within cyberspace can manifest themselves in physical destruction. The Aurora experiment, sponsored by the Department of Homeland Security, demonstrated how effecting the SCADA system used to control one of the most common generators in the electrical power grid could lead to violent destruction.[26] Researchers from Idaho National Laboratories infiltrated the network control system for the generator and provided surreptitious commands that caused it to catastrophically fail. Although the details of the experiment itself are classified, the results are publically available and even published on YouTube.[27] This research reveals the broader impacts that a coordinated attack of this type could have on the electrical power distribution of the U.S.

Unfortunately, the development of SCADA systems incorporated little to no concern for security.[28] As a result, these networks are vulnerable to attack, a risk accentuated by the fact that they form the foundation for controlling most critical infrastructure systems. A SCADA system consists of hardware and software components tailored to a specific task. The system monitors various sensors and provides control through signals to logic controllers, which conduct functions such as opening or closing relays and valves. If an attacker were able to infiltrate this process and alter the status of

---

[26] Jeanne Meserve, "Staged Cyber Attack Reveals Vulnerability in Power Grid," *CNN.com*, 26 September 2007, http://articles.cnn.com/2007-09-26/us/power.at.risk_1_generator-cyber-attack-electric-infrastructure?_s=PM:US (accessed 18 Sep 2011).
[27] Ibid.
[28] Ibid.

the system, actual or perceived, the results could be catastrophic. Because of the unique application of SCADA systems and the importance of their reliability, they are very expensive to procure. In addition, hacking efforts on SCADA systems require direct access to the specialized software and hardware associated with a particular system.[29] As an example, the STUXNET SCADA attack required ownership of the exact hardware and software associated with the system as well as months, if not years, of development and testing of a cyberspace weapon.[30] This reality currently limits the field of would be attackers to nation-states with the resources and expertise to conduct such an attack. These countries are most notably the United States, Russia, China, Israel, India, and North Korea. Although this list will certainly grow, the point is that the field is limited to a group of actors that help scope the current deterrence problem. As a result, the threat of a terrorist group bringing down a nuclear reactor using STUXNET is not significant enough to consider at the moment. However, SCADA devices used by normal civilian industries are more easily obtained and cyber terrorists could conceivably develop and test a computer worm to attack with potentially catastrophic results. Given that threat capabilities will continue to grow, the U.S. assumes greater and greater risk of attack the longer a search for a solution persists.

Both STUXNET and the Aurora experiment are examples of how capabilities within cyberspace can have far-reaching strategic effects. The U.S. is currently too open and vulnerable to attacks of this nature to provide a credible deterrent. Without a viable defense against attacks of this nature, the best the U.S. can hope for is to either strike first or somehow blunt an initial attack and retain a second strike capability.

---

[29] Rose Tsang, "Cyberthreats, Vulnerabilities and Attacks on SCADA Systems," *Berkley.edu* http://gspp.berkeley.edu/iths/Tsang_SCADA%20Attacks.pdf (accessed 29 Dec 2011).
[30] Ibid.

Credible Threats

As identified in the 2010 *National Security Strategy*:

> *The threats we face range from individual criminal hackers to organized criminal groups, from terrorist networks to advanced nation states. Defending against these threats to our security, prosperity, and personal privacy requires networks that are secure, trustworthy, and resilient. Our digital infrastructure, therefore, is a strategic national asset, and protecting it—while safeguarding privacy and civil liberties—is a national security priority.*[31]

The active criminal element operating through cyberspace is perhaps a more worrisome threat in the near term than the full-scale attack of a nation state. Much of the criminal activity within cyberspace is unreported to the public for fear that customer confidence in a specific company will diminish, thus driving away business. An estimated one trillion dollars has been lost to cybercriminals with a very low prosecution rate.[32] General Keith B. Alexander, U.S. Cyber Command (USCYBERCOM) Commander, cited this as the "greatest theft that we have seen in history."[33] There is also increased evidence that nation-states could be harboring or encouraging this activity. Both China and Russia have large criminal organizations operating in cyberspace with suspected state sponsorship. When considering how to combat this threat, General Alexander believes that "we live in a glass house" and must first develop a viable defensive posture before taking the offensive in anything less than a national emergency.[34] Although this defense encompasses government networks (specifically those within the military ".mil" domain), the current strategy does not include those operating outside of this sub-domain of cyberspace. The Department of Homeland

---

[31] U.S. Government, *The National Security Strategy of the United States of America*, 27.
[32] Donna Miles, "Head of U.S. Cyber Command Cites Need for Greater Cyber Defenses," defpro.com, http://www.defpro.com/news/details/27762/ (accessed 18 Sep 2011).
[33] Ibid.
[34] Ibid.

17

Security has taken the responsibility to enforce security of the ".gov" domains. The current philosophy regarding the civilian ".com" domain is one in which it is the responsibility of the company or the individual to secure their own networks and information. This approach does not make sense in light of the far-reaching impacts an outside attack would cause. It would be the equivalent of asking cities to pay for and operate their own missile defense systems to protect themselves from nuclear-armed reentry vehicles during the Cold War. Logically, the government should develop and enforce the appropriate deterrence strategy to enable security procedures and leadership in order to protect the very vulnerable civilian cyber commons. Likewise, there must also be a corresponding understanding of when activity within cyberspace equates to an attack against the U.S. and thus necessitates a military response.

Cyber War

It is also very important and necessary to understand the full scope of possible cyber war scenarios before one can establish effective deterrence strategies. What constitutes "cyber war" in many publications today is more appropriately classified as criminal activity. Assaults on government and civilian networks currently take place thousands of times each day with no significant repercussion to those responsible. Banks lose millions each year to cybercrime, yet few choose to report this activity for fear of affecting the confidence of their customers, a practice known as the "code of silence."[35] Espionage activities within cyberspace yield terabytes of compromised data likely piped directly to the Chinese government, fueling rapid weapons productions efforts. There is increasing evidence that these efforts are also prepping the cyber battlefield through the

---

[35] Clay Wilson, "Cyber Crime," in *Cyberpower and National Security* (Washington D.C.: Potomac Books, 2009), 428.

installation of covert back door programs that more easily enable future attacks.[36] Are these acts of war? These activities do not currently fit any characterization of cyber war principally because no coherent definition or policy exists. This makes deterrence of these types of activities minimal or non-existent because there has not been a consistent record of accomplishment for punishing people, organizations, nations who commit crime or espionage within cyberspace.[37]

This problem has two principle components. First, the U.S. is highly dependent on information systems and easily vulnerable to attack. This means that a compromise of information systems often yields significant payoff to the intruder. A relatively lax view of security within cyberspace, in favor of accessibility, enables this activity. Second, there is no declaratory statement from the U.S. government indicating where the line in the digital sand exists and what the repercussions will be. In short, the unlikely risk of detection or punishment does little to discourage malicious cyberspace activity, while the benefits are immense.

The most recent meeting of the U.S. and Australian defense and diplomatic chiefs focused on adding a cyber warfare clause to the ANZUS treaty among the United States, Australia and New Zealand.[38] This treaty is a first step in defining what constitutes an attack. In addition, such a declaration could establish a collective defense agreement in the context that an attack on one treaty member is an attack on all. This treaty would also define cyber terrorism as an act of war, thus codifying for the first time what cyber war is. With these precepts in mind, a useful definition of cyber war is then any attack

---

[36] Richard A. Clark and Robert K. Knake, *Cyber War* (New York: Harper Collins, 2010), 31.
[37] Ibid.
[38] Kendra Srivastava, "U.S., Australia add Cyber-Warfare to Defense Treaty," MOBILEDIA.com, September 15, 2011, http://www.mobiledia.com/news/108368.html (accessed 15 Sep 11).

through cyberspace, to include espionage activity, that aims to bring harm to the homeland or its populace. There have been very few examples of true war within the cyberspace domain but Russia's attack on Georgia, as discussed above, represents the clearest example up to this point. With the precedent set by the ANZUS treaty, the U.S. and others in larger alliances, like NATO, can now take steps to establish a demarcation line that, if crossed, results in appropriate actions being taken against the perpetrators.

From a deterrence standpoint, the proof is in the follow-on action, which must be severe enough to dissuade possible future attackers for fear of the repercussion. Deputy Defense Secretary William J. Lynn addressed this concept when briefing NATO leaders in Belgium on 15 September 2010 and formalized these thoughts in the *Department of Defense Strategy for Operating in Cyberspace* in July 2011. This new strategy incorporates "active defense," which enables offensive operations against those probing critical networks within cyberspace. General Alexander, reiterated these ideas and indicated that "the government is adopting an active defense strategy aimed at bolstering the readiness of computer networks to respond."[39] General Alexander is referring to the reactive stance that the Department of Defense currently has in defending government networks. Because of the time lapse in detecting network intrusions and then responding, network defenders wait for a successful intrusion to occur and then react to the consequences. Tens of thousands of these attempts take place on government networks each day, thwarted by network defense tools and tactics, techniques, and procedures employed by network defenders. General Alexander's vision for an active defense calls for a near real-time detection of an incursion that would enable the appropriate proactive

---

[39] Bill Gertz, "Computer Based attacks emerge as threat of Future, General Says," *The Washington Times*, September 13, 2011.

20

reaction to stop intruders at the source before damage to U.S. systems occurs.[40] In reality, nations have already taken steps to prepare the battlefield for offensive operations by implanting backdoor code, which provides a gateway into networks that control critical infrastructure, and SCADA systems, in preparation for attack.[41] The most likely culprits are the Chinese, who are constantly developing both offensive and defensive military capabilities utilizing familiar doctrine and integrating cyberspace, although others may not be far behind.[42]

As the next chapter reveals, the history of deterrence in other warfighting domains offers parallels to cyberspace in the form of basic principles that are universal to the primary objective of preventing an attack, and should deterrence fail, ensure victory in conflict. Before examining this history, however, it is useful to analyze existing strategic guidance to determine the gaps and flaws of our current cyberspace strategy.

Strategic Guidance

The renowned strategic theorist Harry R. Yarger defines strategy as; "the calculation of objectives (ends), concepts (ways), and resources (means) within acceptable bounds of risk to create more favorable outcomes than might otherwise exist by chance or at the hand of others."[43] Currently, documents such as the *National Security Strategy*, *National Defense Strategy*, *National Military Strategy*, and *Quadrennial Defense Review* provide strategic ideas with respect to cyberspace. While current strategy defined in these seminal documents emphasizes the importance of deterrence and

---

[40] Bill Gertz, "Computer Based attacks emerge as threat of Future, General Says," *The Washington Times*, September 13, 2011.
[41] Clark, *Cyber War*, 31.
[42] Guocheng Jiang, "Building an Offensive and Defensive PLAAF," *Air and Space Power Journal* 24, no. 2 (Summer 2010): 85.
[43] Harry R. Yarger, *Strategic Theory for the 21st Century: The Little Book on Big Strategy* (Washington DC: Strategic Studies Institute, 2006), 1.

the significance of cyberspace, it does little to tie them together. According to Dr. James Blackwell, noted author and special advisor to the Air Force Chief of Staff, the overall understanding of deterrence in defense circles has atrophied. Current deterrence discussions are mostly associated with either nuclear employment or the movement of forces to seize the initiative during a campaign.[44] In the current environment, the nation would do well to revisit classic deterrence theory.

National Security Strategy

The President's National Security Strategy (NSS) states that "we are pursuing new strategies to protect against biological attacks and challenges to the cyber networks that we depend upon."[45] As mentioned previously, the aim of this thesis is to advance this effort through a historical analysis and application of deterrence strategies. A top priority in the NSS is to "secure cyberspace" by investing in people and technology and developing domestic and international partnerships.[46] This strategy will also "place renewed emphasis on deterrence and prevention by mobilizing diplomatic action, and use development and security sector assistance to build the capacity of at-risk nations and reduce the appeal of violent extremism."[47] Although this deterrence emphasis is not explicitly linked to cyberspace it does reflect an emphasis on the importance of both to national security.

---

[44] James Blackwell, "Deterrence at the Operational Level of War," *Strategic Studies Quarterly* 5, no.2 (Summer 2011): 30.
[45] *National Security Strategy of the United States of America*, 4.
[46] Ibid., 27.
[47] Ibid., 48.

The International Strategy for Cyberspace

The President's *International Strategy for Cyberspace*, published in May 2011, supplements the NSS guidance and provides a self-proclaimed roadmap for future U.S. operations within cyberspace. This document characterizes the cyberspace deterrence strategy of the U.S. as "deterring and dissuading" potential threats.[48] The strategic guidance emphasizes that these efforts will involve collaborating with like-minded nations in order to establish acceptable norms for operating in cyberspace.[49] In addition, this document provides a formal declaration of cyberspace deterrence policy with the following:

> *All states possess an inherent right to self-defense, and we recognize that certain hostile acts conducted through cyberspace could compel actions under the commitments we have with our military treaty partners. We reserve the right to use all necessary means—diplomatic, informational, military, and economic—as appropriate and consistent with applicable international law, in order to defend our Nation, our allies, our partners, and our interests. In so doing, we will exhaust all options before military force whenever we can; will carefully weigh the costs and risks of action against the costs of inaction; and will act in a way that reflects our values and strengthens our legitimacy, seeking broad international support whenever possible.*[50]

There is little evidence, however, to indicate that this new strategy is more than rhetoric. Attacks within cyberspace continue relatively unabated. In addition, strategic guidance from both the Secretary of Defense and Chairman of the Joint Chiefs of Staff does little to expand on the Presidential strategy.

---

[48] U.S. Presidential Publication, "International Strategy for Cyberspace," (May 2011): 13.
[49] Ibid., 9.
[50] Ibid., 14.

Sustaining Global Leadership: Priorities for the 21st Century Defense

Most recently, the DoD published a comprehensive new strategy defined in *Sustaining Global Leadership: Priorities for 21st Century Defense*.[51] The need for the strategic review and renewed guidance for the force was brought on by the end of the war in Iraq, the downsizing of the force waging war in Afghanistan, and the inevitable defense cuts generated by the Budget Control Act of 2011. This document defines "Deter and Defeat Aggression" as a primary military objective using a "combined arms campaign across all domains – land, air, maritime, space and cyberspace" to deter aggressor objectives.[52] This strategy also emphasizes that countries, such as China and Iran, may use asymmetric capabilities, including cyberspace attacks, to challenge the access of the U.S. to the global commons.[53] Another primary objective is to "Operate Effectively in Cyberspace and Space" using "advanced capabilities to defend its networks, operational capability, and resiliency in cyberspace and space."[54] This guidance emphasizes the importance of cyberspace and deterrence but, as mentioned earlier, does little to link the two.

Quadrennial Defense Review

The Secretary of Defense's 2010 *Quadrennial Defense Review* (QDR) specifies that one of the primary objectives of the Defense Department is to "operate effectively in cyberspace."[55] The necessity to "actively defend networks" in order to enable the

---

[51] U.S. Department of Defense, *Sustaining U.S. Global Leadership: Priorities for 21st Century Defense*, 6.
[52] Ibid., 4.
[53] Ibid.
[54] Ibid., 5.
[55] U.S. Department of Defense, *Quarennial Defense Review Report* (Washington DC: Department of Defense, Feb 2010), ix.

24

necessary access to cyberspace is a key component of DoD operations.[56] The QDR calls

for both centralized command of cyber operations as well as enhanced partnership to

achieve this goal. Deterring conflict while advancing the common interests of the U.S.

and its allies hinges on the capabilities of diplomacy, development, defense, intelligence,

law enforcement, economic tools, and statecraft all working in concert.[57] The QDR calls

for a "tailored deterrence" approach that focuses less on military influence and more on

understanding capabilities, values, intent, and decision making in order to use all

instruments of power to deter aggression from both state and non-state actors.[58] This

approach is more co-option than deterrence in that is tries to motivate potential threat

actors through more of a "carrot and stick" approach similar to that described with North

Korea and the Seven Party Talks. The DoD also plans to strengthen cyberspace

capabilities by establishing a comprehensive approach to operations in cyberspace,

developing greater cyberspace expertise and awareness, centralizing command under

USCYBERCOM, and enhancing partnerships with other agencies and governments.[59]

These efforts focus primarily on building up the defenses on DoD networks to prevent

attacks from achieving their desired effect. Absent from both the co-option and heavy

network defense based deterrence approaches is the threat of repercussion. Co-option

through "tailored deterrence," a strong defense, and a credible offense are all necessary

components of a cyberspace deterrence strategy. The QDR does not fully highlight the

interplay between these three essential requirements.

---

[56] U.S. Department of Defense, *Quarennial Defense Review Report*, ix.
[57] Ibid., 13.
[58] Ibid., 14.
[59] Ibid., 38.

Department of Defense Strategy for Operating in Cyberspace

This DoD Strategy for Operating in Cyberspace supplements the QDR and provides a more comprehensive approach to operations within cyberspace.[60] The Defense Department defines five broad initiatives with respect to cyberspace:[61]

1. Treating cyberspace as an operational domain
2. Employing "active defense" to protect DoD networks and systems
3. Partner government agencies and private sector to enable whole-of-government approach
4. Build robust relationships with allies and international partners
5. Leverage ingenuity through exceptional workforce and innovation

Deterrence of threats within cyberspace is noticeably absent from this list but employing "active defense," a capability deemed essential by USCYBERCOM, does suggest a step toward a policy with more repercussions for those who probe, infiltrate or exploit DoD networks. The focus on partnerships and a "whole-of-government approach" is indeed crucial to the success in deterring cyberspace threats, but there is no clear distinction of who will lead the charge.

National Military Strategy

The Chairman of the Joint Chiefs of Staff highlights in the *National Military Strategy (NMS) of the United States of America* that "cybersecurity threats represent one of the most serious national security, public safety, and economic challenges we face as a nation."[62] Despite this pervasive threat, there is no further expansion in the NMS of efforts to deter cyberspace threats. Moreover, Joint doctrine on cyberspace operations and deterrence in general does not exist. The *Deterrence Operations Joint Operating*

---

[60] U.S. Department of Defense, *Department of Defense Strategy for Operarting in Cyberspace* (Washington DC: Department of Defense, July 2011).

[61] Ibid., 1.

[62] U.S. Joint Chiefs of Staff, *The National Military Strategy of the United States of America* (Washington DC: Joint Chiefs of Staff, Feb 8 2011), 27.

Concept (DO JOC) published by U.S. Strategic Command in 2006 remains the only formalized deterrence guidance and represents a consolidation of deterrence practices and planning considerations.[63] The DO JOC emphasizes key deterrence concepts and ideas, but does not distill them into a broader deterrence strategy. The same holds true with the most recent guidance provided by the Department of Homeland Security.

Blueprint for a Secure Cyber Future

In November 2011, the Department of Homeland Security (DHS) published the *Blueprint for a Secure Cyber Future* designed to create a framework for a "safe, secure and resilient cyber environment."[64] The primary theme is to increase resiliency of the "cyber ecosystem" by accomplishing 20 separate objectives supporting the following eight goals:

1. Reduce Exposure to Cyber Risk
2. Ensure Priority Response and Recovery
3. Maintain Shared Situational Awareness
4. Increase Resilience
5. Empower Individuals and Organizations to Operate Securely
6. Make and Use More Trustworthy Cyber Protocols, Products, Services, Configurations and Architectures
7. Build Collaborative Communities
8. Establish Transparent Processes

These goals, tied to defense of the "cyber ecosystem," seemingly provide little more than political rhetoric for an ideal environment with no direct correlation to or basis in reality. In other words, the ways and means to accomplish these ends are speculative and poorly defined. In addition, the use of a "cyber ecosystem," rather than a shared definition of cyberspace as the domain of reference, strains the coordination between DHS and DoD.

---

[63] U.S. Strategic Command, *Deterrence Operations Joint Operating Concept, Version 2.0.* (Omaha: U.S. Strategic Command, December 2006).

[64] U.S. Department of Homeland Security, *Blueprint for a Secure Cyber Future* (Washington DC: Homeland Security, November 2011), ii.

Lack of a common lexicon is a strong symptom of communication breakdowns between DHS and DoD, further indicating failure of these organizations to properly coordinate broader cyberspace deterrence efforts. Stove-piped organizations are easily constructed and breaking down these communication barriers is an incredible challenge. Cyberspace deterrence requires unprecedented information sharing and harmonization to be effective and DHS and DoD must be the hallmark for leadership and coordination in this effort.

Many factors are influencing cyberspace deterrence, and the habitual inaction of the U.S. is a sign of weakness that may be cause for an adversary's adventurism. The next chapter provides a historical analysis of deterrence. This effort will emphasize key principles that highlight important lessons learned from the past in an effort to rectify current shortfalls and pave the way to effective deterrence within cyberspace in the future.

CHAPTER 2: HISTORY

> *History does not repeat itself—but it does rhyme.*
>
> - Mark Twain

The somewhat repetitive flow of history, while not precise or exact in its cycles or events, is nonetheless familiar. History unfolds from era to era in familiar patterns because man, the one constant, does not change his fundamental nature. When Clausewitz writes that "war is a human endeavor," he warns modern warriors that, while the tools and the geography of war may change, the fundamental reasons men (or nations) fight or not remain unchanged through time. By this same logic, the nature and overarching principles that guide warfare are timeless regardless of the domain. As an example, some regard cyberspace as a unique domain that no longer fits within the traditional paradigms. Cyberspace does offer distinctive capabilities and asymmetric avenues of attack that make its capabilities and employment unique in some applications. However, the same basic principles of waging war within cyberspace remain unchanged when compared to other warfighting domains. The real decision U.S. policymakers must make, is whether to confront challenges in cyberspace with these principles in mind or patch them into developing policy later. The difference is crucial, but the failure of government and military professionals to recognize and incorporate lessons of history at the outset is not a new phenomenon.

Thinking in Time

In 1986, Richard E. Neustadt and Ernest R. May, two distinguished Harvard University professors, published *Thinking in Time*. This book formalized a process that they had been teaching for years at Harvard's John F. Kennedy School of Government to compensate for the lack of general history knowledge among military and government decision makers.[1] Starting in the 1970s, Neustadt and May taught "Uses of History" to countless midcareer professionals including legislators, ambassadors and senior military officers. They started the course because they identified a noticeable lack of knowledge about history and an all-to-frequent problem of viewing contemporary problems as new and fresh, rather than recognizing the historic parallels.[2] Over the years, they developed a system that helped these same high-level decision makers recognize and capitalize on history in their daily lives.

The methodology developed by Neustadt and May addresses and applies to the current unfocused and disjointed "stabs" at cyberspace deterrence strategy development, described in the previous chapter. By using history to supply analogies that best fit the circumstances, problem framing becomes more effective and efficient, while solutions manifest themselves more readily. The collection of analogous similarities and differences help in applying what they call the "Goldberg Rule" which is simply asking "what is the story?"[3] By telling the story through the lens of history, the problem itself becomes more refined and solutions tied to previous meaningful observations begin to materialize. This process is what the authors call "thinking in time."

The remainder of this chapter will highlight historical examples of deterrence in order to identify the similarities and differences in the analogies they create with cyberspace. Although this seems to be an elementary approach to understanding and

---

[1] Richard E. Neustadt and Ernest R. May, *Thinking in Time*, (New York: The Free Press, 1986), 244.
[2] Ibid., xi.
[3] Ibid., 244.

solving problems, failure to capitalize effectively on historical analogies continues as a key limiting factor today in critical analysis. This process establishes trends across history with respect to deterrence that coalesce into fundamental principles for developing an effective and efficient deterrence strategy.

Because deterrence is a fundamental aspect of waging war, the principles associated with deterrence are universal to all domains, including cyberspace. A closer examination of deterrence in a broad historical context shows how established basic principles will best shape a deterrence strategy for cyberspace.

Space Deterrence

The Chinese demonstrated the vulnerability of space-based assets with the destruction of a retired weather satellite using an anti-satellite missile in 2007.[4] This demonstration alarmed many strategists as it exposed limitations in the ability to deter such aggression. This concern led to a focus on building satellite system resilience as well as diplomatic negotiations aimed at limiting offensive actions in space. The cost of space systems, and weapons to target them, currently limit negotiations to China, Russia and the United States.

The U.S. has operated with impunity within the space domain since the launch of the first military satellites. Despite the current vital importance of the space domain and the dependence of military systems on space borne hardware, relatively little thought has gone into space defense or deterrence. The U.S. continues to operate with virtually complete freedom, but as more countries develop capabilities that threaten space-based

---

[4] Craig Covalt, "Chinese Test Anti-Satellite Weapon," *Aviation Week & Space Technology,* January 17, 2007, http://www.aviationweek.com/aw/generic/story_channel.jsp?channel=space&id=news/CHI01177.xml (accesed November 2, 2011).

assets, the mindset must change. The most likely deterrence strategy would involve limiting development of antisatellite capabilities and policing threats to space through cooperative security agreements and treaties while simultaneously using technology to defend these key national assets.[5]

The 1967 Outer Space Treaty precludes participating nations, including the United States, from positioning weapons of mass destruction for launch from outer space.[6] This treaty does not preclude the deployment of other types of weapons in space, yet the U.S. has not taken overt steps to arm space primarily because of the precedent that it would set. Arming space would likely lead to unwelcome consequences since both Russia and China would respond in kind, thus initiating a space-based arms race. The same issues arise with regard to cyberspace capabilities as continued employment of advanced offensive capabilities will set a precedent that could ultimately lead to prolific development of advanced network attack weapons. Arms limitation treaties will not likely be an effective tool for deterring attacks within cyberspace, however. Computer viruses, worms, and other attack tools are simply too easy to conceal and, therefore, arms limitation agreements would be impossible to enforce, much like the deterrence efforts associated with weapons of mass destruction.

Deterrence of Weapons of Mass Destruction (WMD)

As Derek Smith indicates in the book *Deterring America*, the U.S. has significantly downgraded the nuclear threat over the past two decades. Following the

---

[5] Barry R. Posen, "Command of the Commons: The Military Foundation of U.S. Hegemony," *International Security 28*, no. 1 (Summer, 2003): 14.

[6] Unites States Department of State, TREATY ON PRINCIPLES GOVERNING THE ACTIVITIES OF STATES IN THE EXPLORATION AND USE OF OUTER SPACE, INCLUDING THE MOON AND OTHER CELESTIAL BODIES, http://www.state.gov/www/global/arms/treaties/space1.html (accesed November 2, 2011).

Cold War, the threat of chemical, biological, radiological and nuclear (CBRN) weapons employed by either terrorist organizations or "rogue states" is now taking center stage.[7] The successful deterrence of the use of these weapons both in Iraq and Libya offer valuable lessons learned for the subsequent effort in cyberspace.

Iraq

It was widely understood that Iraq possessed Weapons of Mass Destruction and a propensity to use them prior to and during the 1990-91 Gulf War. United Nations inspections after Desert Storm confirmed the deployment of both chemical and biological weapons in the rear echelons and in defense of Baghdad prior to the U.S. invasion.[8] Saddam Hussein never ordered the use of these weapons, however, and his most likely calculus hinged on the strong likelihood of an overwhelming response from the United States when compared with the uncertain consequences of his own WMD. In a March 2003 interview on NBC's "Meet the Press," then Vice President Dick Cheney stated, "we've always adopted the policy that if someone were to use a weapon of mass destruction—chemical, biological or nuclear—against the United States or U.S. forces, we reserve the right to use any means at our disposal to respond."[9] Although Saddam Hussein may have reserved these weapons in the event that coalition forces invaded Baghdad, the more likely estimate is that the repercussions outweighed the reward and the deterrence efforts of the U.S. were a success. The U.S. was well equipped to deter Iraq in part because of the equipment to protect troops from chemical and biological attacks as well as the diplomatic and military power to punish Iraq severely for their

---

[7] Derek D. Smith, *Deterring America: Rogue States and the Proliferation of Weapons of Mass Destruction* (Cambridge: University Press, 2006), 50-51.

[8] Ibid.

[9] Vice President Dick Cheney, interview with Tim Russert, *Meet the Press*, NBC, March 16, 2003, http://www.mtholyoke.edu/acad/intrel/bush/cheneymeetthepress.htm (accessed February 20, 2012).

actions. Just as they do today, troops had been trained and certified in utilizing the robust

CBRN protection and decontamination equipment in combat and support operations.

This not only acted as a deterrent, but theoretically enabled forces to fight through an

attack if deterrence failed.

These lessons are applicable to the cyberspace defensive posture of the U.S. as

well. The collective training and certification of personnel using networks is the best

frontline defense against attack.[10] These efforts have been somewhat successful, but

must improve ahead of and commensurate with the likely increase in the frequency and

sophistication of attacks. As discussed previously, the effectiveness with which users

protect themselves from attack weighs in the attacker's decision matrix. A decrease in

the likelihood of an effective attack acts as a deterrent as does the likelihood of detection.

Fear of detection, coupled with the likelihood of the attack to escalate to a conventional

or even nuclear response, serves as one of the strongest deterrence capabilities within

cyberspace. These same principles applied to deterring aggression from Libya in the mid

1980s.

Libya

In April 1986, the U.S. conducted air strikes on Libyan targets within the cities of

Tripoli and Benghazi.[11] Operation EL DORADO CANYON was the military response to

years of Libyan sponsored terrorist attacks that included airport bombings in Rome and

Vienna as well as the bombing of the La Belle discotheque in West Berlin that killed two

---

[10] U.S. Department of Homeland Security, *Blueprint for a Secure Cyber Future* (Washington DC: Homeland Security, November 2011), 20.

[11] Craig R. Black, *Deterring Libya: The Strategic Culture of Muammar Qaddafi* (Montgomery: Air War College, 2000), 7.

American service members and wounded 79 others.[12] This military strike marked the escalation of a deterrence campaign focused on Libyan leader Muammar Qaddafi. This strategy was eventually effective in stopping Qaddafi's overt sponsorship of terrorism as well as ending his aggressive pursuit of the development, proliferation and use of WMD. After EL DORADO CANYON, Qaddafi immediately ceased sponsorship for terrorist bombings against American targets. In 1992, in response to nearly a decade of economic sanctions that crippled the Libyan economy, Qaddafi began negotiations with the U.S. that would eventually lead to him abandoning his WMD pursuits.[13] These efforts not only deterred the use of WMD, but also eventually led to Qaddafi's overthrow during the 2011 Arab Spring revolution. As with the case of Iraq, there are strong parallels with this snapshot of history and deterring attacks within cyberspace. Any deterrence efforts will require strong diplomacy as well as significant consequences, both military and economic, in order to succeed. The bombing of a near-peer threat, such as China, in response to an unauthorized network intrusion involves much larger ramification than those experienced following El DORADO CANYON. Given this reality, the U.S. must make a firm commitment on what constitutes a cyberspace attack as well as determining what the appropriate military response should be. In addition, the U.S. must have a credible defensive capability that can allow for the inevitability of attack, as mentioned above. A good defense should allow an operation to be successful by enabling forces to fight through cyberspace threats.

---

[12] Bruce Jentelson and Christopher Whytock, "Who 'Won' Libya? The Force-Diplomacy Debate and Its Implications for Theory and Policy," *International Security*, 30, no 3 (Winter 2005): 58.
[13] Ibid., 18.

Iran

The Arab Spring and political power play in South West Asia presents an incredible deterrence challenge with countries such as Iran. Decades of efforts to balance the powerful Iranian regime has achieved varied levels of success. Iran's nuclear arms capability, although limited by sanctions and efforts such as STUXNET, seems to be inevitable. Ironically, Iran looks at America's handling of Libya and Iraq as an example of why it must pursue nuclear weapons and not succumb to the deterrence efforts of the United States.[14] Libya capitulated to the deterrence strategy of the U.S. and chose to abandon efforts to develop nuclear weapons. The ultimate result was that the United States, as part of a larger NATO effort, assisted in the recent overthrow of the Qaddafi regime. Iraq begrudgingly revealed details of its own WMD capability for fear of showing weakness to neighboring Iran as well as internal enemies of the Hussein regime.[15] Paradoxically, this move ultimately led to military action and regime change during Operation IRAQI FREEDOM.[16] The likelihood of regime change in Libya or Iraq would have been immeasurably less had these countries possessed nuclear weapons.[17]

North Korea is a perfect example of the enabling capability of nuclear weapons to a regime. Kim Jong-Il remained in power largely because of the nuclear capability that North Korea possesses.[18] The same is true in the near term as the country transitions to new leadership under Kim Jong-Un. Any external threat to the North Korean regime would almost certainly result in the real possibility of a nuclear exchange. Some question if having a single nuclear weapon or even a small number is enough to provide the

---

[14] Austin G. Long, *DETERRENCE: From Cold War to Long War* (Washington D.C.: RAND, 2008), 74-75.
[15] Ibid.
[16] Ibid.
[17] Ibid.
[18] Ibid.

appropriate deterrent. The reality is that there is no such thing as a limited nuclear war. Research indicates and history validates the truth that the destructive power of a single nuclear weapon can inflict enough damage to deter most conflicts.[19] As long as an adversary has the capability to inflict unacceptable damage, can protect that capability from a first strike, and can credibly communicate intentions to use the weapon, then deterrence will succeed.[20] Therefore, the number of nuclear weapons is not the key to deterrence, but rather the capability to achieve these three fundamental elements. The current Iranian leadership seems to understand this concept as well as the approach the U.S. has taken to prevent the proliferation of nuclear weapons. Iran likely sees itself as succumbing to the same regime change as Libya if it agrees to the pleas from the West to refrain from adding nuclear weapons to their arsenal. The history of U.S. assistance in regime change thus shapes the paradigm of Iranian President Mahmoud Ahmadinejad as he contemplates the terms proposed through diplomatic negotiations with the United States. Although Ahmadinejad is President, Supreme Leader Ali Hoseini-Khamenei is the actual ruler of the country. This leadership dynamic complicates the deterrence focus as Khamenei has the ultimate veto power and may or may not agree with the Iranian President.[21] The success of the Arab Awakening in inciting regime change makes the ambition of both the President and the Supreme Leader that much more desperate to demonstrate the power to preserve the Iranian establishment. In effect, the approach used to deter nuclear weapons in Libya now weakens the capability to do the same with Iran. This knowledge of the precedent set with deterrence efforts throughout history

---

[19] John Lewis Gaddis, *Strategies of Containment : A Critical Appraisal of Postwar American National Security* (London: Oxford University Press, 1982), 273.
[20] Ibid.
[21] Long, *DETERRENCE: From Cold War to Long War*, 75.

demonstrates the imperative to incorporate lessons from the past in current deterrence efforts. In keeping with this idea, the U.S. must be cognizant of current and future cyberspace deterrence efforts and realize the danger of a "one size fits all" approach.

Nuclear Deterrence

Deterrence in cyberspace can also benefit from the evolution of the combined arms approach associated with the Kennedy administration's "Flexible Response" strategy during the 1960's. Flexible Response advanced U.S. military capability by shifting the deterrence strategy away from President Eisenhower's nuclear dominated "New Look." Rather than assuming that an automatic and all-out response would be required if nuclear deterrence failed, Flexible Response "would place a number of retaliatory options in the hand of the President so that responses to attacks could be rational, deliberate and controlled."[22] In concert with this approach, the U.S. military developed a more balanced structure that included nuclear, conventional and special operations forces.[23] Despite these advances, Flexible Response continued to rely on the New Look strategy of "an explicit nuclear retaliatory threat" in order to counter the Soviet intimidation of Europe.[24] This dependence on nuclear weapons justified a ground force that was not large enough to deter the U.S.S.R. by conventional means, leaving a nuclear onslaught the only outcome if the two great nations went to war.[25]

Similar patterns exist today in identifying a deterrence strategy in cyberspace in that offensive capabilities alone (like nuclear weapons) will not deter actors within cyberspace. A leaner force more reliant on attacking through cyberspace could fall

---

[22] David W. Tarr, *American Strategy in the Nuclear Age* (New York: MacMillan Publishing, 1966), 69.
[23] Ibid., 99.
[24] Ibid.
[25] Ibid.

victim to the same pitfalls as described above. By focusing on the similarities in this

analogy, a better approach would be to understand that reliance on a single type of

military capability generates incredible risk. As mentioned previously, forces must be

capable of fighting through a cyberspace attack in order to deliver the necessary

counterattack. In addition, an adversary must understand that this capability exists for it

to act as a deterrent.

The Cuban Missile Crisis

The Cuban Missile Crisis represents thirteen very long days in October 1962 that

stand as one of the most crucial and breath-taking events in deterrence theory and

American history. A moment when two countries faced mutually assured destruction

from nuclear war, but paused and reconciled because two leaders were able to

communicate effectively in a time of crisis. The fact that the Cuban Missile Crisis did

not culminate in a nuclear exchange finds it roots in the discourse between the leadership

of the U.S. and the Soviet Union at the time. By establishing a naval "quarantine" of

Soviet ships steaming toward Cuba with nuclear missiles intended for use against the

U.S., President Kennedy was able to demonstrate American resolve to escalate the crisis

to the point of war.[26] In addition, President Kennedy's message during his public address

to the nation on October 22, 1962 focused deliberately on the leader of the Union of

Soviet Socialist Republics, Chairman Nikita Khrushchev:

> *I call upon Chairman Khrushchev to halt and eliminate this clandestine,*
> *reckless, and provocative threat to world peace and to stable relations*
> *between our two nations. I call upon him further to abandon this course*
> *of world domination, and to join in an historic effort to end the perilous*
> *arms race and to transform the history of man. He has an opportunity now*

---

[26] Graham Allison and Philip Zelikow, *Essence of Decision* (New York: Addison-Wesley, 1999), 122.

to move the world back from the abyss of destruction -- by returning to his government's own words that it had no need to station missiles outside its own territory, and withdrawing these weapons from Cuba -- by refraining from any action which will widen or deepen the present crisis -- and then by participating in a search for peaceful and permanent solutions.[27]

President Kennedy also wrote Chairman Khrushchev prior to the public address, outlining that, "I have not assumed that you or any other sane man would in this nuclear age, deliberately plunge the world into war which it is crystal clear no country could win and which could only result in catastrophic consequences to the whole world, including the aggressor."[28] After realizing he had underestimated the U.S. response, Chairman Khrushchev became willing to negotiate, ultimately trading removal of the U.S. Jupiter nuclear missiles in Turkey for the U.S.S.R. missiles in Cuba.[29] With respect to cyberspace, the U.S. must also have a clear leadership focus and open dialog when communicating the deterrence message. In addition, the U.S. should have the resolve to escalate to conventional or even nuclear weapons employment depending on the severity of the cyberspace attack. These two elements of classic deterrence theory are lacking in the current deterrence strategy for cyberspace.

Military Deterrence

Mere military deterrence in general has had limited effect throughout history and the size of an army seems to do little to prevent the actions of a determined adversary.[30] Although you cannot wage war effectively without utilizing a full spectrum approach to conflict (all warfighting domains), land war is historically the only decisive means of victory. In this respect, the logic of Clausewitz is timeless, indicating that "war is a clash

---

[27] John F. Kennedy, public address October 22, 1962, *The John F. Kennedy Presidential Library,* http://microsites.jfklibrary.org/cmc/oct22/doc5.html (accessed February 21, 2012).
[28] Ibid.
[29] Graham Allison and Philip Zelikow, *Essence of Decision,* 351.
[30] Ibid., 329.

between two major interests, which is resolved by bloodshed—that is the only way in which it differs from other conflicts."[31] If two men confront one another in a bar there are any number of ways that they can convey their intentions to one another, but eventually someone is going to either back down or a fight will ensue. The same is true of nations led by men. How does this relate to cyberspace? There seem to be no scenarios observed or imagined which indicate that cyberspace breaks the above paradigm. David Lonsdale in *The Nature of War in the Information Age* rightfully concludes that physical violence is simply a part of war.[32] To prevent this reality, a common deterrence strategy is protection in numbers, otherwise known as collective defense.

In short, the more credible and supportive friends a person, or nation, has the less likely someone will be to engage in violence against them (survival instinct). The outcome of World War I created the infant stages of an effective deterrence tool in the League of Nations. The concept that a collective defense would be more effective than acting alone developed into the principles of the United Nations following World War II. In 1951, the North Atlantic Treaty Organization adopted the concept of collective defense summarized in Article 5 of the treaty, which states that an attack on one treaty member is an attack on all.[33] The implications for cyberspace are more complex, however, because of such fundamental yet contentious issues like defining what truly constitutes an

---

[31] Carl Von Clausewitz, *On War*, (New York: Random House, 1993), Book 2, Chap 3, 173.

[32] David J. Lonsdale, *The Nature of War in the Information Age*, (New York: Frank Cass, 2004), 207.

[33] North Atlantic Treaty Organization, "Agreement Between Parties of NATO Regarding the Status of their Forces," 15 June, 1951, http://www.nato.int/cps/en/natolive/topics_59378.htm?selectedLocale=en (accessed 13 Feb 12).

attack.[34] The inability to reach a consensus among NATO members prevented the invocation of Article 5 in 2007 when Russia attacked Estonia, a member since 2004.[35] This massive cyberspace onslaught crippled both government and civilian information systems and cost the Estonian government millions of dollars, yet no overt action was taken against the Russian government.[36] This issue has yet to be resolved still today.

Up to this point, the historical examination of cyberspace deterrence has focused primarily on key underlying similarities with other domains. Neustadt and May emphasize the importance of looking at both similarities and differences while thinking in time. Interestingly, Martin Libicki highlighted some of the most relevant differences in his book *Cyberdeterrence and Cyberwar*. Unfortunately, due to his widespread acclaim as a cyberspace expert for the RAND Cooperation, Libicki has created a paradigm within the U.S. government focused on the differences rather than the similarities between cyberwar and warfare. This approach has promulgated the idea, a misguided understanding, that the problem of cyberspace deterrence is new and fresh, therefore, isolating many experts knowledgeable in deterrence, while less informed about cyberspace capabilities. This mindset has resulted in the current strategic guidance that misses the mark in providing an effective deterrence strategy. The next section will dissect Libicki's arguments and highlight the importance of history in creating an effective cyberspace deterrence strategy, addressing many of the questions he posits.

---

[34] Haly Laasme, "Estonia: Cyber Window into the Future of NATO," *Joint Forces Quaterly* 63, no. 4 (October 2011): 60.
[35] Laasme, "Estonia: Cyber Window into the Future of NATO," 60.
[36] Ibid.

How is Cyberspace Different?

Richard Clarke, National Coordinator for Security, Infrastructure Protection and Counterterrorism for President Clinton, clearly demonstrates a Libicki-like bias in his book, *Cyber War*. As an example, Clark proposes additional layers of bureaucracy, like a Cyber Defense Administration and national security investigatory center to focus effort on this "new" problem.[37] Yet, the U.S. is clearly less in need of additional layers of command and technical deterrence solutions than it is for a coherent game plan to deter both direct and indirect threats within cyberspace. Although cyberspace is a new domain, as outlined above, deterring threats is familiar territory. Despite this fact, Libicki asserts that the following questions do not have adequate answers to support a strategy for deterrence within cyberspace:[38]

> Who did it?
> Can their assets be repeatedly held at risk?
> If deterrence fails, what about disarmament?
> Will third parties join the fight?
> Does retaliation send the right message?
> What is the threshold for response?
> Can escalation be avoided?
> What if the attacker has little worth hitting?

It should be readily apparent to the reader that Libicki's questions are not unique to operations in cyberspace. Based on the historical similarities described earlier, the answers to these same questions arise in applying deterrence to any domain. Although more obvious in areas such as nuclear weapons employment, these questions mirror the considerations for deterrence strategies throughout the ages. They are interestingly a culmination of the key analogous differences associated with cyberspace deterrence. The following section will address this in more detail.

---

[37] Clark, *Cyber War*, 265.
[38] Martin C. Libicki, *Cyberdeterrence and Cyberwar*, (California: RAND, 2011), 39-70.

Attribution

Who did it? Much like the deterrence of threats possessing WMD, cyberspace attackers are difficult to detect making retaliation in cyberspace a complex problem.[39] By using the parallels with deterrence of WMD, there are key steps that aid in the deterrence of threats within cyberspace. Much like the use of WMD, unless the attack is an overt strike from a state actor, a robust investigation of the source will likely be required prior to targeting. The credibility, speed, and success of this investigation will lead to future deterrence of similar attacks. The key is to demonstrate the capability to identify attackers and prosecute them either legally or militarily in a relatively short amount of time. General Alexander suggests that an automated response to an attack through cyberspace could augment the idea of active defense, but with current technological capabilities, the ambiguity behind the source of the attack makes this a high-risk approach.[40] Thus reinforcing current technological capabilities is a key strategy analogous to the continuous technological improvements that occurred in nuclear technology during the Cold War.

Can their assets be repeatedly held at risk? This question highlights the necessity for a full spectrum response to cyberspace attacks. As with the development of the Flexible Response strategy during the Cold War, each time an attack takes place, the appropriate and proportional military response should be commensurate with the seriousness of the event. This is not a graduated response, but rather a realization that an act of war in cyberspace constitutes retaliation by all means available. The conventional

---

[39] Eric Sterner, "Retaliatory Deterrence in Cyberspace," *Strategic Studies Quarterly* 5, no. 1 (Spring 2011): 62.

[40] Donna Miles, "Head of U.S. Cyber Command Cites Need for Greater Cyber Defenses," defpro.com, http://www.defpro.com/news/details/27762/ (accessed 18 Sep 2011).

deployment or buildup of forces will likely be necessary to demonstrate the resolve of the U.S. to punish those who choose to attack. This is also analogous to the WMD deterrence strategy used with both Libya and Iraq. Current strategic guidance indicates that this is the strategy of the U.S. with respect to cyberspace attacks but it is heretofore unobserved.[41]

Disarmament

If deterrence fails, what about disarmament? If deterrence fails then the response must assure deterrence in the future. By demonstrating the commitment to punish attackers, history shows that the U.S. will be far more likely to deter future action. Disarmament is not logically feasible for two primary reasons. First, cyberspace is not largely a military domain, but rather wholly integrated within the fabric of the civilian and business arenas. Secondly, the capabilities within cyberspace are much too easy to conceal from the outside world. Although treaties have been historically effective in downsizing the nuclear arsenal, it is much more difficult to ensure an offensive cyberspace weapon is no longer a viable threat. The judicious implementation of space-based weapons, as described above, offers the best analogy to cyberspace. This could be applied successfully to cyberspace by limiting the availability and possible use of offensive cyberspace weapons. This precedent will likely deter the aggressive pursuit of these weapons by other state and non-state actors. In addition, collective defense agreements, like ANZUS, will help enforce acceptable behavior of like-minded nations to coerce aggressors to limit cyberspace as a viable asymmetric attack avenue. However, bilateral treaties with countries promising not to employ cyberspace attack weapons

---

[41] U.S. Presidential Publication, "International Strategy for Cyberspace," 13.

should not be trusted. The ability to hide such weapons nearly guarantees that such a treaty would be ineffective.

Retaliation

Does retaliation send the right message? The answer to this question, much like any other, is dependent on the situation. Bombing Libyan targets in 1986 was appropriate retaliation and augmented the broader deterrence strategy of the United States. How can this apply to cyberspace? According to the *International Strategy for Cyberspace*, each time China attacks the U.S. through cyberspace could generate the appropriate conventional military buildup to demonstrate the commitment to defend national sovereignty.[42] Each time China threatens the sovereignty of Taiwan, the U.S. likely responds by deploying naval assets to the region, at a minimum. The sovereignty of cyberspace needs to receive the same level of attention. Failure to do so results in a lack of credibility in acting on the line drawn by the *International Strategy for Cyberspace*. Just as with any other military response, costs, options and the likelihood of success must be carefully weighed.

Escalation

What is the threshold for response? This is fundamental to any deterrence strategy. A declaratory statement must state where the line in the sand is and how the U.S. will respond. As previously discussed, the *International Strategy to Secure Cyberspace* indicates that attacks on the U.S. through cyberspace warrant a full range of military response options. The key component of this message is enforcement. If attacks continue unabated, the credibility of this deterrence message is lost. As described above,

---

[42] U.S. Presidential Publication, "International Strategy for Cyberspace," (May 2011): 13.

one key to preventing Saddam Hussein from using WMD in Operation DESERT STORM was a firm belief that the U.S. would use nuclear weapons in response.[43]

Can escalation be avoided? History shows that, given the correct approach to deterrence, the U.S. can avoid escalation. The lessons of communication and transparency gleaned from the Cuban missile crisis are just as applicable to cyberspace. The *International Strategy for Cyberspace* states that the United States will "reserve the right to use all necessary means—diplomatic, informational, military, and economic—as appropriate and consistent with applicable international law, in order to defend our Nation, our allies, our partners, and our interests."[44] Given enforcement of this deterrence message, with a clear focus on the decision maker, escalation can be avoided.

Targeting

What if the attacker has little worth hitting? The best offense is not necessarily the best defense with respect to cyberspace. As discussed in deterrence of WMD, cyberspace defense must be sufficient to convince any aggressor that attack will result in failure to achieve the desired effects. Even if the U.S. could invoke a STUXNET style attack on a country's power generation capability, there is the persistent concern that the same thing can be done in return. As described earlier, these strategic cyberspace weapons take years to develop and are designed against specific systems. A country may not have a cyberspace capability worth targeting, thus requiring a more conventional attack through air, land or sea domains. An attack must occur nonetheless but the response does not need to come through cyberspace. A full spectrum military reaction, as

---

[43] Long, *DETERRENCE*, 74-75.
[44] U.S. Presidential Publication, "International Strategy for Cyberspace," (May 2011): 13.

dictated in the *National Military Strategy*, is the best deterrent and mitigates the lack of potential corresponding targets within cyberspace.

Toward a Deterrence Strategy for Cyberspace

Time is still on the side of the U.S. in attempting to solve the cyberspace deterrence problem. The significance of cyberspace threats and the need for deterrence is beginning to permeate discussions at the highest levels within the U.S. government. Unfortunately, the majority of these discussions are ill-informed, misplaced or the threats are exaggerated. The primary reason for this is that a basic understanding of the cyberspace environment is unclear to many Americans, including our senior leaders, and the door remains open for "cyber experts" to fill the void. The most damaging course of action would be for the U.S. government to leap forward without first understanding the environment and the problem. Unfortunately, this has been an all-to-common approach up to this point. The military, and government as a whole, is failing to deter malicious actors within cyberspace. Near term, there must be significant steps to secure the networks of government systems and those that control the nation's critical infrastructure. The U.S. is hemorrhaging sensitive information to hackers from various countries and the solution will be somewhat disruptive and costly. The first step is defining this activity as an act of war. The U.S. cannot take that step however, without the resolve to back it up.

The next chapter focuses the search for a deterrence solution. Using basic principles gleaned from the examination of the similarities and differences described above, the intent is to focus deterrence strategy in order to avoid past failures and capitalize on current successes as well as technological advances.

CHAPTER 3: THE PRINCIPLES OF CYBERSPACE DETERRENCE

> *I would not give a fig for the simplicity this side of complexity, but I would give my right arm for the simplicity on the other side of complexity.*
>
> - Oliver Wendell Holmes

The majority of current thinking with respect to deterrence of threats within cyberspace is that the best scenario possible is one of Mutually Assured Destruction.[1] There is tremendous incentive in safeguarding this critical domain, but there is currently no capability to do so.[2] The search for a dazzling technical solution will likely never end as the highly skilled hackers of the world continue to launch wave upon wave of attacks against government and civilian networks. Hackers of all origins introduce over 55,000 pieces of malicious computer software, or "malware," each day, over one every two seconds.[3] Yet, the biggest concern to General Alexander and his staff at USCYBERCOM is the more destructive element that is likely on the horizon.[4] Private industry has incentive in developing hardware and software to prevent intrusion by malware and other attacks because industry and government agencies continue to purchase these programs in search of a cure-all. Unfortunately, the rate of malware development and the prolific nature of hackers means that the advance of anti-virus software and firewall technology will at best keep pace depending on the amount of money and effort dedicated to the problem. The U.S. government can follow a path to an

---

[1] Matthew D. Crosston, "World Gone Cyber MAD: How 'Mutually Assured Debilitation' is the Best Hope for Cyber Deterrence," *Strategic Studies Quarterly* 5, no. 1 (Spring 2011): 100.
[2] Martin C. Libicki, *Cyberdeterrence and Cyberwar*, 1.
[3] Donna Miles, "Head of U.S. Cyber Command Cites Need for Greater Cyber Defenses."
[4] Ibid.

effective cyberspace deterrence strategy by examining key historical approaches and extracting key principles.

Deterrence Principles

Because human nature does not change much over time, the psychological aspects of deterrence do not change either. Given that deterrence fundamentally deals with the human decision-making process, the development of a deterrence strategy in one domain has strong parallels in another. If the definition of cyber war developed in the ANZUS treaty holds true, then the U.S. is already at war in cyberspace. In other words, deterrence has failed. The good news is that the current war in cyberspace has not reached a threshold in which bullets, bombs, and missiles have begun to fly. Cyberspace battles will continue, and with more frequency and severity, until a sufficient deterrence strategy emerges. The following principles, derived from an analysis of selected historical periods, aim to advance this effort.

Principle 1: Define the Domain

Private companies manage much of what comprises the Internet with somewhat arbitrarily applied access rules. Countries like China and Russia have used censorship with limited success in an attempt to force the boundaries of the Internet to conform to their own national boundaries. China's efforts collided with private industry recently when government hackers infiltrated Google's private Gmail user account database and utilized the information to target suspected dissidents.[5] In response, Secretary of State Hillary Clinton issued a demarche against the Chinese government and Google shut down

---

[5] Mark Landler and Edward Wong, "China Rebuffs Clinton on Internet Warning," *The New York Times*, January 22, 2010.

service to China.[6] The U.S. Government has made it very clear that initiatives to secure cyberspace, including the Internet, will not limit the "openness and interoperability that have characterized its explosive growth."[7] Defining the domain means describing what type of activity is prohibited in cyberspace.

Much like the use of WMD, unless the attack is an overt strike from a state actor, military planners must thoroughly investigate the source prior to targeting. The credibility, speed, and success of this investigation will deter similar attacks in the future. The key is to demonstrate the capability to identify attackers and prosecute them either legally or militarily depending on the situation. Preventing espionage requires the credible threat of violence. Crime prevention necessitates the threat of prosecution. Deterring nation state attacks requires the intimidation of an escalatory response with a full range of offensive capabilities and culminating in the most severe kinetic retaliation, nuclear weapons.

The U.S. military cannot expect to dominate cyberspace. A strategy of selective engagement will help to scope the domain and provide a more reasonable approach to maintaining a dominant global posture. Selective engagement represents the capability to dominate with military force in those areas of the global commons, including cyberspace, crucial to achieving national objectives. Command of the commons is the key military enabler of the U.S. global power position.[8] Cyberspace is a critical military and domestic domain that expands the global commons in yet another dimension. Defense of the

---

[6] Mark Landler, "China Rebuffs Clinton on Internet Warning."

[7] U.S. Presidential Publication. "International Strategy for Cyberspace." (May 2011): 3.

[8] Barry R. Posen, "Command of the Commons: The Military Foundation of U.S. Hegemony," *International Security* 28, no. 1 (Summer, 2003): 8.

crucial cyberspace infrastructure and data is just as important as maintaining airspace or sea-lanes to both the U.S. and global economies.

Principle 2: Defend the Domain

The U.S. was well equipped to deter Iraq, in part, because of the equipment to protect troops from chemical and biological attacks as well as the diplomatic and military power to punish Iraq for their actions. These lessons are applicable to cyberspace as well. The limited success of effectively training individual users to detect and prevent attacks on military networks must improve based on the likely increase in the frequency and sophistication of attacks.

The current strategy dominated by defense is doomed to failure because it is impossible to defend everything.[9] The U.S. military must establish priorities for the defense of those cyberspace capabilities that enable combat forces to sustain an initial attack and launch a counter attack. The essential elements of a cyberspace offense include the power grid, the internet fiber optic backbone, and military networks. These systems enable combat capability within cyberspace and the U.S. must put a top priority on their defense. Although unthinkable in the near term, this defensive strategy will likely involve removing these critical systems and networks from the internet and/or adopting a new and more secure protocol.[10]

A key lesson from both Iraq and Libyan deterrence models is that military power alone did not deter aggression, but hinged on a lengthy diplomatic effort coupled with economic sanctions. Any deterrence effort will require strong diplomacy as well as

---

[9] Richard Stiennon, *Surviving CYBER WAR*, (United Kingdom: Government Institutes, 2010), 154.
[10] Rajesh Talpade, "Toward Foolproof IP Network Configuration Assessments," in *Cyber Infrastructure Protection* (Carlisle: Strategic Studies Institute, 2011), 263.

repercussions, both military and economic, in order to be successful. As described in Chapter 2, the past deterrence strategy of the U.S. against other state actors can set a precedent that may preclude the same approach in the future. Iran will likely become a nuclear power in order to prevent the same fate of the Libyan regime. The U.S. must have a credible defense that allows for the inevitability of a cyberspace attack. In addition, the U.S. is rapidly approaching a point where the vulnerabilities associated with the internet surpass the enabling capabilities. Given this reality, there must be the ability to fight through a cyberspace attack with the accompanying training of forces prior to war.

Principle 3: Destroy Threats to the Domain

George Washington stated in his fifth address to Congress on the 3rd of December 1793, "If we desire to avoid insult, we must be able to repel it; if we desire to secure the peace, one of the most powerful instruments of our rising prosperity, it must be known that we are at all times ready for war."[11] To deter threats within cyberspace, the U.S. must guarantee a credible threat of destruction and a clear commitment to punish any aggressor. In the nuclear arms race, a clear declaratory statement by the U.S. removed any doubt in the adversary's mind that full scale nuclear war (and mutually assured destruction) would result from an attack on the American homeland. As described in Chapter 1, the U.S. has declared the intent to use all manner of force to defend itself from attack, including those initiated through cyberspace. The problem arises when there is a failure to act. If a network attack is an act of cyberwar then opponents must see a history of positive measures that back up the words in the

---

[11] George Washington, Congessional Address, December 3, 1793, http://stateoftheunion.onetwothree.net/texts/17931203.html (accessed February 21, 2012).

International Strategy for Cyberspace. There must also be an element of transparency to the capabilities possessed by the U.S. so that any potential adversary understands what to expect if they choose to launch an attack. Military strikes on Libya in 1986 clearly demonstrated the resolve of the U.S. and shaped the paradigm of Libyan leader Muammar Qaddafi in his nuclear weapons pursuits. Both state and non-state actors, such as Al-Qaeda, can be effectively deterred in the same manner through a simple, well articulated and broadly communicated deterrence message with a proven record of enforcement. It must also be clear that the U.S. will hold countries responsible for what goes on within their borders in order to prevent safe havens for criminal and terrorist organizations that currently use cyberspace with relative impunity.

Principle 4: Beware of Treaties

The absence of weapons on space borne assets provides an example of how treaties have been effective in limiting arms buildup, but also identifies another important element of deterrence--restraint. Restraint limited widespread weapons proliferation in space far more than the 1967 Outer Space Treaty. By simply limiting the development of space based weapon systems, the U.S. prevented an arms race in which other countries would have been forced to match capabilities in order to deter an American attack. Similarly, because the presence of cyberspace weapons is difficult to verify, the U.S. must rely less on formal arms limitation treaties and instead promote restraint in the use of offensive cyberspace capabilities. Wide spread employment of advanced offensive capabilities will set a precedent that could ultimately lead to prolific development of advanced network attack weapons. The employment of STUXNET against Iran presents an excellent example. In employing STUXNET, the creator(s) may have inadvertently placed a weapon in the hands of an enemy that they had no hope of producing on their own. In an interview for NBC's "60 Minutes," Sean McGurk, the Director of Control System Security for DHS, equated the employment of STUXNET to opening Pandora's Box.[12] McGurk stated "you can download the actual source code of STUXNET now and you can reproduce it and repackage it and then, you know, point it back towards wherever it came from."[13]

Principle 5: Establish Escalation Precedence

Secretary Cheney left little doubt that Iraq's use of WMD would warrant a nuclear response. It is worthwhile to compare crisis management and the interplay of cyber weapons and strategy with that of nuclear weapons.[14] The most significant lesson from the Cold War era is the resolve to escalate a conflict. The deterrence message should be clear and overt to avoid any confusion regarding the resolve of the U.S. to act if attacked. The *International Strategy to Secure Cyberspace* accomplishes this requirement, but U.S. threats will have little credibility without a commitment to act on the deterrence message and punish attackers. In fact, employment of nuclear weapons should be weighed in the response options following a cyberspace attack on the United States.

Principle 6: Ensure a Flexible Response

The U.S. should be cognizant of the historic lessons of Eisenhower's "New Look." In the current resource constrained environment, a significantly downsized ground force with strong dependence on cyberspace capabilities will likely have the same

---

[12] Sean McGurk, interview by Steve Kroft, *60 Minutes*, NBC, March 4, 2012.
[13] Ibid.
[14] Stephen J. Cimbala, "Nuclear Crisis Management and 'Cyberwar': Phishing for Trouble?" *Strategic Studies Quarterly* 5, no. 1 (Spring 2011): 117.

limitations in deterring conflicts as that seen during the 1950s leading up to the Vietnam War. The concept of Flexible Response is just as applicable today as it was decades ago. The likelihood is that offensive cyberspace capabilities alone will not discourage aggressors, necessitating a more flexible array of full spectrum capabilities to deter threats. Forces must be prepared to fight through a cyberspace attack and respond in the most appropriate manner. General Alexander's concept of active defense has the potential to evolve into such a posture, in which unauthorized cyberspace intrusions are neutralized at the source before they can cause harm to U.S. systems. The use of such offensive employment through cyberspace is authorized under section 954 of the *National Defense Authorization Act for Fiscal Year* 2012.[15]

Principle 7: Institute a Collective Defense

The ANZUS treaty is the first step in defining what constitutes a cyberspace attack as well as a mindset that an attack on one treaty member constitutes an attack on all members, no matter what the domain or avenue of attack. The precedent set by the ANZUS treaty will likely enable great strides in limiting the likelihood of attack through cyberspace. This offers the most promise for establishing an environment both free of criminal activity as well as mitigating the risk of a crippling military attack. Treaties, like NATO, that provide collective defense have huge promise in deterring cyberspace threats. As the current strategy indicates, the U.S. must continue to build cooperative agreements with like-minded governments, but cautiously approach treaties that limit offensive or defensive capabilities. In addition, the effectiveness of treaties against non-

---

[15] United States Congress, *The National Defense Authorization Act for Fiscal Year 2012* (Washington D.C.: Congressional Record, 12 Dec 2011), H.R. 1540, Volume 157, Number 190, H8356-8726.

state actors or terrorists depends on the ability to hold countries who host these organizations responsible for what goes on within their borders.

Principle 8: Demilitarize Foreign Policy

Potential adversaries are well aware of the ability of the U.S. to operate in a WMD environment. The development of equipment and robust training, exercises and inspections continually validate unit readiness in this mission area. Although this factors into the decision to attack American forces, the homeland remains at significant risk to WMD attack. The proliferation of WMD and the inability to uncover every potential threat leaves the U.S. with the reality that prevention of war through deterrence strategies is the only recourse. The unrivaled military capability of the U.S. cannot target all WMD capability and, as demonstrated in the second war in Iraq, determining the existence of WMD capability may itself be a formidable task.[16] With these challenges in mind, the U.S. should focus more on a strategy to defeat adversaries when required, but also to develop a more peaceful international environment by focusing on diplomacy and "demilitarizing foreign policy."[17] As the terminology would indicate, demilitarizing foreign policy involves every effort to utilize non-military instruments of national power in an effort to encourage other countries to do the same. The military is used only as a last resort in resolving conflicts. This strategy is useful with respect to cyberspace for many of the same reasons. By choosing to use cyberspace as another domain of war, the U.S. invites that same activity on its own networks by adversaries. The simple act of

---

[16] Austin G. Long, *DETERRENCE: From Cold War to Long War* (Washington D.C.: RAND, 2008), 74-75.

[17] Mr. Y, *A National Strategic Narrative* (Washington D.C.: Woodrow Wilson Center, 2011), 3.

limiting offensive military actions within cyberspace will likely have a calming influence on those who may be otherwise inclined to attack.

Principle 9: Determine the Focus of the Deterrence Effort

A clear understanding of the object of deterrence efforts is paramount. Deterrence efforts in Libya and Iraq indicate that any strategy must have a clear leadership focus and a simple declaratory message to avoid the errors of the past. This is much more difficult when considering an attack initiated through cyberspace. Who is the offensive organization's decision maker and how is the deterrence message to be communicated? Identification of the key decision maker is crucial so that deterrence efforts focus simply and clearly on this individual or group. An adversary must understand that the U.S. has the capability to launch an attack and to protect itself from being attacked. This can only be done through careful advertisement directed at the opposing leadership, as demonstrated by the successful resolution of the Cuban Missile Crisis.

Principle 10: Continually Incorporate History

As the above nine principles demonstrate, history does provide a foundation upon which to build an effective deterrence strategy. Cyberspace deterrence is a different problem, not a new problem. By harnessing the *Thinking in Time* methodology and heeding the similarities and differences in historic analogies, strategic planners and decision makers develop a more effective and efficient cyberspace deterrence solution.

CHAPTER 4: RECOMMENDATIONS

> *For want of a nail the shoe was lost. For want of a shoe the horse was lost. For want of a horse the rider was lost. For want of a rider the battle was lost. For want of a battle the kingdom was lost. And all for the want of a horseshoe nail.*
>
> - George Herbert

This chapter highlights the future initiatives required to move the U.S. along the path to a more robust deterrence posture within cyberspace.

Principles of Cyberspace Deterrence

The principles defined in Chapter 3 need to be socialized within USCYBERCOM, DHS, and DoD to ensure incorporation into strategic guidance and decision making. The fact that cyberspace does not change the nature of warfare is not widely accepted and the basic deterrence principles will help break this paradigm. This is just the beginning in developing a coherent cyberspace deterrence strategy, however. Neustadt and May's "Goldberg Rule" requires that further refinement of these principles by asking the "when, what, where, who, how and why."[1] The idea is to have one coherent strategy between both DoD and DHS which encompasses these principles and tells the story so that effective use of the limited means available may be positioned correctly to help prevent a devastating cyber attack. In short, "the more history one knows, the better one understands the options."[2]

Joint Doctrine Development

Joint doctrine for deterrence or for operations within cyberspace does not currently exist. Service level efforts in the Army and Air Force provide an excellent

---

[1] Richard E. Neustadt and Ernest R. May, *Thinking in Time*, (New York: The Free Press, 1986), 241.

[2] Ibid., 245.

starting point, but the joint community must have common guidance for the employment of forces. The Army Training and Doctrine Command has an exceptional framework for joint doctrine development in *Cyberspace Operations Capability Plan 2016-2028*.[3] In addition, the Air Force AFDD 3-12, *Cyberspace Operations*, established the basic relationship of cyberspace capabilities, the principles of war, and tenets of air and space power. Finally, the integration of military deception techniques into deterrence strategy development provides key insights for influencing the decision making of the adversary.[4]

Developing and Communicating the Strategy

Paramount to developing a coherent deterrence strategy is the communication of that strategy to potential adversaries. This can come through demonstration of capabilities through exercises or more overtly through doctrine or treaties, such as ANZUS. The latter efforts take the form of Department of State diplomatic cables that must be synchronized among agencies such as DHS and DoD to ensure that a concise and clear message makes its way to any potential adversaries. This is effectively drawing the line in the sand that will trigger an escalated response intended to deter adversary action. Military engagement, security cooperation and deterrence underlie the full range of military operations. Deterrence should be the key element of American foreign policy with military force used as a last resort. The focal point of the deterrence strategy must be the leadership of the country or organization in question and the utmost effort put into clearly communicating America's intentions along with the associated incentives and

---

[3] U.S. Army, *Cyberspace Operations Concept Capability Plan 2016-2028*, TRADOC Pamphlet 525-7-8 (Virginia.:U.S. Army Training and Doctrine Command, 22 February 2010), 1.
[4] Debra K. Rose, "Only in the Mind of the Enemy: Can Effectiveness be Measured?" (Master's Thesis, Joint Advanced Warfighting School, 2011), 1.

consequences resulting from positive or negative behavior. The friction associated with communication and decision-making requires the utmost simplicity and clarity.

Legal Approval

There are tremendous legal hurdles to traverse in order to employ both kinetic and non-kinetic weapons in response to a cyberspace attack.[5] The U.S. cannot ignore the legal aspects of strategic employment and warfighting in cyberspace.[6] The laborious legal approval process cannot hamper the rapid employment of offensive cyberspace capabilities--a critical element of an effective deterrence strategy. There must be a way to preapprove targets for a response given certain parameters.

Training and Exercises

The development of realistic training venues offers the unique capability to train forces across the range of military operations. Cyberspace operates within and across this same range of operations and encompasses Joint and interagency personnel in a cooperative stance to both defend and destroy assets within the domain. The development of strategic, operational, and tactical level exercises for cyberspace forces is the key to a unified effort. The Cyber Flag exercise sponsored by USCYBERCOM in October 2011 is just such an environment. Knowledge of this type of training can in fact advance deterrence efforts. In short, the enemy must know that we are training, who we are training with, and why. In his historic analysis of warfare Raoul Naroll concluded that, "it is often said that God is on the side of the largest battalions. If God takes sides, it is not the side of the largest battalions, nor that of the best fortified battalions, nor that of

---

[5] Matthew C. Waxman, "Cyber-Attacks and the Use of Force: Back to the Future of Article 2(4)," *The Yale Journal of International Law*, 36, (April 2011): 421-423.

[6] Maj Gen Charles J. Dunlap Jr., USAF, Retired, "Perspectives for Cyber Strategists on Law for Cyberwar," *Strategic Studies Quarterly* 5, no. 1 (Spring 2011): 81.

the most renowned. Rather God, if he takes sides at all, seems to stand at the side of the well trained battalions."[7]

Historically, it is to the advantage of all involved if the decision is made to avoid conflict altogether. Exercises are an advertisement for how prepared the U.S. is to respond militarily and for whom we plan to have on our side when the shooting starts. Training and exercises are more fundamental to warfighting in cyberspace than in any other domain. This results from the requirement to coordinate both government and civilian agencies to defend the domain and to fight through any projected attack. The U.S. is more vulnerable to an asymmetric attack than to a conventional attack due to the extent of the dependence on technology. Training and exercises offer a way to not only develop expertise within cyberspace, but also a strong deterrence posture in that adversaries understand that the U.S. has the capability to take and hold the high ground in cyberspace while denying such action on its own military and civilian systems. Planning for war in cyberspace must be a joint and interagency effort because of the ubiquitous nature of this domain. A recent GAO report (released 20 Jun 11) criticizes the services for a lack of uniformity in preparing for both offensive and defensive operations in cyberspace.[8] If the ultimate goal is to develop both offensive and defensive capabilities to deter aggression, then there needs to be solid doctrinal guidance that solidifies the roles of each agency and then realistic training and exercises to test it. Although the Nation is defending itself every day against small-scale threats to the network, the only way to establish mainstream capabilities for a full-scale cyber war is to train as if we would

[7] Raoul Naroll, Vern L. Bullough and Frada Naroll, *Military Deterrence in History* (New York: State University of New York Press, 1974), 337.

[8] Staff Writer, "GAO report: No uniform approach to cybersecurity training," *Air Force Times*, July 4, 2011.

expect to fight. With a firm understanding of the threats and their capabilities, the U.S. must establish a training construct that allows forces to defend against attackers as well as launch offensive operations. Knowledge of this type of training will inherently serve as a deterrent to potential adversary as well as develop sharp network attackers and defenders. The time is right for Cyber Flag.[9]

Command and Control

As Hayden indicates in his book *Warfighting*, "our philosophy of command must be based on human characteristics rather than on equipment and procedures."[10] There are currently no campaign examples, seen or postulated, where cyberspace can be a decisive domain. It is much like air and space in this regard, irrespective of nuclear weapons delivered through these domains. The current trend to equate cyberspace to something new and different is at the root of the current dysfunctional Command and Control (C2) architecture used to control cyberspace operations. Because cyberspace operations most closely parallel space operations, the current C2 architecture resembles that of the space domain. As Major General Williams surmised during his time as Director of Command, Control, Communications and Computer Systems for U.S. Pacific Command, "we should apply existing capstone doctrinal tenets regarding strategy, operational art, and C2 relationships to cyberspace operations, and our analogy should be the terrestrial domain."[11] Warfighting in the cyberspace domain presents difficulties for the Combatant

---

[9] Andrew P. Hansen, "Cyber Flag: A Realistic Training Environment for the Future" (Master's Thesis, Air Force Institute of Technology, 2008), 1.

[10] Tom Hayden, *Warfighting: Manuever Warfare in the U.S. Marine Corps* (London: Greenhill Books, May 1995), 1.

[11] Brett T. Williams, "Ten Propositions Regarding Cyberspace Operations," *Joint Forces Quarterly*, no. 61 (2d Quarter 2011): 12.

Commanders and the command and control of forces.[12] These difficulties manifested themselves in the Schriever Wargame 2010.[13] For example, manning does not currently exist to support each combatant commander with the appropriate forces for offensive and defensive actions in cyberspace. As a result, forces operate from a consolidated location under USCYBERCOM. The result is an architecture in which the geographic responsibilities and functional responsibilities bump into one another rather than overlapping. The solution is rooted in the development of a cyberspace Joint Operating Area (JOA) controlled by the combatant commander.[14] Although borders of cyberspace do not abide by the geographic boundaries set for the combatant commanders in the Unified Command Plan, there must be a capability for unity of command of all forces tied to a Joint Task Force.

[12] Maj M. Bodine Birdwell and Lt Col Robert Mills, "War Fighting in Cyberspace," *Air and Space Power Journal 25, no. 1 (Spring 2011): 26.*

[13] Brig Gen Terrence J. O'Shaughnessy, Lt Col Baron V. Greenhouse and Lt Col Kurt M. Schendzielos, "Effects Felt Around the World: The Growing Complexities of the Interaction Between Geographic and Functional Combatant Commanders," *High Frontier* 7, no. 1 (November 2010): 30.

[14] Ibid.

CONCLUSION

What has been will be again, what has been done will be done again; there is nothing new under the sun. Is there anything of which one can say, "Look! This is something new?" It was here already, long ago; it was here before our time.

- Ecclesiastes 1:9-10

Deterrence in general is lacking in the strategic planning of the United States. Cyberspace has renewed the importance of a strategic vision, but the strategy should encompass all warfighting domains and implement all instruments of power. The rippling effects of a firm and effective deterrence strategy should find itself at the core of contemporary planning efforts.

History is the foundation for the art of deterrence. As Moltke states, "history is the most effective way to teach war during peacetime." Sub-state actors can wield more power in cyberspace at a lower entry cost than in other domains. However, this power does not directly correlate to successful employment against the United States. With strong offensive capability and an even stronger defensive posture, the U.S. can successfully achieve cyberspace deterrence.

Currently, there are too many random and isolated good ideas of how to deter malicious activity in cyberspace. A room filled with professional musicians all playing their own music at the same time results in noise! The solution is a coordinated effort to get all of the musicians playing off the same sheet of music, one that includes deterrence principles, while preventing "group think" and preserving the chance for breakthrough thinking. Peace is the ultimate objective of any good warfighting strategy. The key to keeping the peace is foreseeing, planning, and deterring threats rather than waiting for the otherwise inevitable crisis or Pearl Harbor event. Deterrence and a good defense are the best possible investments against the much higher costs of war. The U.S. must ensure that penalties for malicious activity become the norm not the exception. Strength in defense and, in turn, deterrence capability will enable other instruments of power, such as diplomacy to take root. There is time and history is the guide to a solution. The development of crippling cyberspace attacks on a grand scale has yet to emerge. Luckily, there is nothing new under the sun!

BIBLIOGRAPHY

Beeker, Kevin R. 2009. Strategic Deterrence in Cyberspace: Practical Application. Master's Thesis. Air Force Institute of Technology.

Birdwell, M. Bodine and Robert Mills, "War Fighting in Cyberspace." *Air and Space Power Journal 25, no. 1 (Spring 2011): 26.*

Black, Craig R. *Deterring Libya: The Strategic Culture of Muammar Qaddafi.* Montgomery: Air War College, 2000.

Alexander, Keith B. "Building a New Command in Cyberspace." *Strategic Studies Quarterly* 5, no. 2 (Summer 2011): 3.

Chumer, Michael J. "Survivability of the Internet," in *Cyber Infrastructure Protection,* 29. Carlisle: Strategic Studies Institute, 2011.

Cimbala, Stephen J. *Deterrence and Nuclear Proliferation in the Twenty-First Century.* Westport: Praeger Publishers, 2001.

_____. "Nuclear Crisis Management and 'Cyberwar': Phishing for Trouble?" *Strategic Studies Quarterly* 5, no. 1 (Spring 2011): 117.

Clark, Richard A. and Robert K. Knake, *Cyber War.* New York: Harper Collins, 2010.

Clausewitz, Carl V. *On War.* Edited and translated by Michael Howard and Peter Paret. New York: Random House, 1993.

Cohen, Ariel and Robert E. Hamilton. *The Russian Military and the Georgia War: Lessons and Implications.* Carlisle: Strategic Studies Institute, 2011.

Crosston, Matthew D. "World Gone Cyber MAD: How 'Mutually Assured Debilitation' is the Best Hope for Cyber Deterrence." *Strategic Studies Quarterly* 5, no. 1 (Spring 2011): 100.

Davis, Lynn Ethridge. *Limited Nuclear Options: Deterrence and the New American Doctrine.* London: The International Institute for Strategic Studies, 1976.

Demchak, Chris C. and Peter Dombrowski. "Rise of a Cybered Westphalian Age." *Strategic Studies Quarterly* 5, no. 1 (Spring 2011): 32.

Douhet, Giulio. *The Command of the Air.* Translated by Dino Ferrari. New York: Coward-McCann, Inc., 1942.

Dunlap Charles J. "Perspectives for Cyber Strategists on Law for Cyberwar." *Strategic Studies Quarterly* 5, no. 1 (Spring 2011): 81.

Edwards, A.J. *Nuclear Weapons, the Balance of Terror, the Quest for Peace.* London: MacMillan, 1986.

Franz, Timothy. "The Cyber Warfare Professional." *Air and Space Power Journal,* 25, no. 2 (Summer 2011): 87.

Gaddis, John Lewis Gaddis. *Strategies of Containment : A Critical Appraisal of Postwar American National Security.* London: Oxford University Press, 1982.

Gelbstein, Eduardo and Ahmed Kamal. *Information Security.* New York: United Nations Institute for Training and Research, 2002.

Gray, Colin S. *Maintaining Effective Deterence.* Carlisle: Strategic Studies Institute, 2003.

_____. "War-Continuity in Change, and Change in Continuity." *Parameters* (Summer 2010): 5-13.

Hansen, Andrew P. 2008. Cyber Flag: A Realistic Training Environment for the Future. Master's Thesis. Air Force Institute of Technology.

Hayden, Michael V. "The Future of Things Cyber." *Strategic Studies Quarterly* 5, no. 1 (Spring 2011): 3.

Hanovich, Robert L. 2010. Fuzzy Deterrence. Master's Thesis. Joint Advanced Warfighting School.

Jiang, Guocheng. "Building an Offensive and Defensive PLAAF." *Air and Space Power Journal* 24, no. 2 (Summer 2010): 85.

Kahn, Herman. *Thinking About the Unthinkable in the 1980s.* New York: Simon and Schuster, 1984.

Kelley, Olen L. 2008. Cyberspace Domain: A Warfighting Substantiated Operational Environment Imperative. Master's Thesis. U.S. Army War College.

Kenny, Anthony. *The Logic of Deterrence.* Chicago: The University of Chicago Press, 1985.

Krepinevich, Andrew F. *7 Deadly Scenarios.* New York: Bantam Dell, 2009.

Kugler, Richard L. "Deterrence of Cyber Attacks," in *Cyberpower and National Security,* 309-340. Washington DC: Potomac Books, 2009.

Laasme, Haly. "Estonia: Cyber Window into the Future of NATO." *Joint Forces Quaterly* 63, no. 4 (October 2011): 60.

Larkin, Sean P. "The Limits of Tailored Deterrence." *Joint Forces Quaterly* 63, no. 4 (October 2011): 53.

Libicki, Martin C. *Cyberdeterrence and Cyberwar*. California: RAND, 2011.

_____. "Cyberwar as a Confidence Game." *Strategic Studies Quarterly* 5, no. 1 (Spring 2011): 132.

Long, Austin G. *DETERRENCE: From Cold War to Long War*. Washington D.C.: RAND, 2008.

Lonsdale, David J. *The Nature of War in the Information Age*. New York: Frank Cass, 2004.

Loo, Fook Weng. "Decisive Battle, Victory and Revolution in Military Affairs." *The Journal of Strategic Studies 32*, no. 2 (April 2009): 189-211.

Lynn, William J. III. "Lynn Explains U.S. Cybersecurity Strategy." *AmeriForce Military News*. September 16, 2010. http://www.ameriforce.net/news/?tag=deputy-defense-secretary-william-j-lynn-iii (accessed August 23, 2011).

Miller, Robert A. "Cyber war and the dangers of preemption." *International Journal of Critical Infrastructure Protection*, January 19, 2011. http://www.sciencedirect.com/science?_ob=MImg&_imagekey=B8JGJ-5207B77-11&_cdi=43672&_user=10&_pii=S1874548211000023&_origin=browse&_zone=rslt_list_item&_coverDate=04%2F30%2F2011&_sk=999959998&wchp=dGLbVlW-zSkWW&md5=b97d5b382161d691008d2fe1798c889e&ie=/sdarticle.pdf (accessed August 9, 2011).

Miller, Robert A., Daniel T. Kuehl, and Irving Lachow. "Cyber War," *Joint Forces Quarterly* 61, no. 2 (2011): 18.

McAfee Foundstone Professional Services and McAfee Labs. "Global Energy Attacks: Night Dragon." *White Paper* (February 2011): 3-4, 18.

Morgan, Patrick M. *Deterrence*. London: SAGE Publications, 1977.

_____. "Applicability of Traditional Deterrence Concepts and Theory to the Cyber Realm." *Proceedings of a Workshop on Deterring Cyberattacks*, 59-76. Washington D.C.:National Academy Press, 2010.

Naroll, Raoul, Vern L. Bullough and Frada Naroll. *Military Deterrence in History*. New
     York: State University of New York Press, 1974.

Neustadt, Richard E. and Ernest R. May. *Thinking in Time*. New York: The Free Press,
     1986.

O'Shaughnessy, Terrence J., Baron V. Greenhouse and Kurt M. Schendzielos, "Effects
     Felt Around the World: The Growing Complexities of the Interaction Between
     Geographic and Functional Combatant Commanders." *High Frontier* 7, no. 1
     (November 2010): 30.

Posen, Barry R. "Command of the Commons: The Military Foundation of U.S.
     Hegemony." *International Security 28*, no. 1 (Summer, 2003): 5-46.

Reule, Fred J. *Dynamic Stability: A New Concept for Deterrence*. Maxwell AFB: Air
     University Press, 1987.

Romjue, John L. *From Active Defense to AirLand Battle: The Development of Army
     Doctrine, 1973-1982*. Fort Monroe, VA: Historical Office, United States Army
     Training and Doctrine Command, 1984.

Rose, Debra K. 2011. Only in the Mind of the Enemy: Can Effectiveness be Measured?
     Master's Thesis. Joint Advanced Warfighting School.

Sheldon, John B. "Deciphering Cyberpower." *Strategic Studies Quarterly* 5, no. 1
     (Spring 2011): 95.

Smith, Derek D. Deterring America: Rogue States and the Proliferation of Weapons of
     Mass Destruction. Cambridge: University Press, 2006.

Starr, Stuart H. "Developing a Theory of Cyberpower." in *Cyber Infrastructure
     Protection*, 15. Carlisle: Strategic Studies Institute, 2011.

Sterner, Eric. "Retaliatory Deterrence in Cyberspace." *Strategic Studies Quarterly* 5, no.
     1 (Spring 2011): 62.

Talapade, Rajesh. "Toward Foolproof IP Network Configuration Assessments," in *Cyber
     Infrastructure Protection*, 263. Carlisle: Strategic Studies Institute, 2011.

Tarr, David W. *American Strategy in the Nuclear Age*. New York: MacMillan
     Publishing, 1966.

Trias, Eric D. and Bryan M. Bell. "Cyber This, Cyber That...So What?" *Air and Space
     Power Journal* 24, no. 1 (Spring 2010): 90.

Tzu, Sun. *The Art of War.* Translated by Samuel B. Griffith. London: Oxford University Press, 1963.

U.S. Air Force. *Air Force Basic Doctrine.* Air Force Doctrine Document 1. Washington DC: Headquarters Air Force, November 17, 2003.

U.S. Air Force. *Cyberspace Operations.* Air Force Doctrine Document 3-12. Washington DC: Headquarters Air Force, July 15, 2010.

U.S. Army. *Cyberspace Operations Concept Capability Plan 2016-2028.* TRADOC Pamphlet 525-7-8. Virginia: U.S. Army Training and Doctrine Command, February 22, 2010.

U.S. Department of Defense. *The Definition of "Cyberspace,"* Gordon England. Memorandum, Deputy Secretary of Defense. Washington D.C., 2008.

_____. *Department of Defense Strategy for Operarting in Cyberspace.* Washington DC: Department of Defense, July 2011.

_____. *Quadrennial Defense Review Report.* Washington DC: Department of Defense, February 2010.

_____. *Sustaining U.S. Global Leadership: Priorities for 21st Century Defense,* 5 January 2012.

U.S. Department of Homeland Security, *Blueprint for a Secure Cyber Future.* Washington DC: Homeland Security, November 2011.

U.S. President. *The National Security Strategy of the United States of America.* Washington DC: Government Printing Office, May 2010.

U.S. Presidential Publication. "International Strategy for Cyberspace." (May 2011): 25.

_____. "The Comprehensive National Cybersecurity Initiatve." (March 2010): 1.

U.S. Joint Chiefs of Staff. *Doctrine for the Armed forces of the United States,* Joint Publication 1. Washington D.C.: Joint Chiefs of Staff, March 20, 2009.

_____. *The National Military Strategy of the United States of America.* Washington DC: Joint Chiefs of Staff, February 8, 2011.

_____. *Department of Defense Dictionary of Military and Associated Terms.* Joint Publication 1-02. Washington DC: Joint Chiefs of Staff, October 2011.

_____. *Information Operations.* Joint Publication 3-13. Washington DC: Joint Chiefs of Staff, February 2006.

U.S. Strategic Command. *Deterrence Operations Joint Operating Concept*. Version 2.0. Omaha: U.S. Strategic Command, December 2006.

Waxman, Matthew C. "Cyber-Attacks and the Use of Force: Back to the Future of Article 2(4)." *The Yale Journal of International Law* 36, (April 2011): 421-423.

Welch, Larry D. "Cyberspace-The Fifth Operational Domain." *IDA Research Notes*. Summer 2011. https://www.ida.org/upload/research%20notes/researchnotessummer2011.pdf (accessed August 8 2011).

Williams, Brett T. "Ten Propositions Regarding Cyberspace Operations." *Joint Forces Quarterly*, no. 61 (2d Quarter 2011): 12.

Yarger, Harry. *STRATEGIC THEORY FOR THE 21ST CENTURY: THE LITTLE BOOK ON BIG STRATEGY*. Washington DC: Strategic Studies Institute, 2006.

Yunzhu, Yao. "China's Perspective on Nuclear Deterence." *Air and Space Power Journal* 24, no. 1 (Spring 2010): 27.

www.ingramcontent.com/pod-product-compliance
Lightning Source LLC
Chambersburg PA
CBHW081847170526
45167CB00007B/2919